NCRP REPORT No. 84

General Concepts for the Dosimetry of Internally Deposited Radionuclides

Recommendations of the
NATIONAL COUNCIL ON RADIATION PROTECTION AND MEASUREMENTS

Issued September 30, 1985

National Council on Radiation Protection and Measurements
7910 WOODMONT AVENUE / BETHESDA, MD. 20814

LEGAL NOTICE

This report was prepared by the National Council on Radiation Protection and Measurements (NCRP). The Council strives to provide accurate, complete and useful information in its reports. However, neither the NCRP, the members of NCRP, other persons contributing to or assisting in the preparation of this report, nor any person acting on the behalf of any of these parties (a) makes any warranty or representation, express or implied, with respect to the accuracy, completeness or usefulness of the information contained in this report, or that the use of any information, method or process disclosed in this report may not infringe on privately owned rights; or (b) assumes any liability with respect to the use of, or for damages resulting from the use of, any information method or process disclosed in this report.

Library of Congress Cataloging in Publication Data

National Council on Radiation Protection and Measurements.
General concepts for dosimetry of internally deposited radionuclides.

(NCRP) report ; no. 84)
Bibliography: p.
Includes index.
1. Radiation—Dosage. 2. Radioisotopes—Migration. 3. Radioisotopes—Physiological effect. 4. Radioisotopes in the body. 5. Radiation dosimetry. 6. Environmental health. I. Title. II. Series.
RA569.N353 1985 616.9′897′00212 85-8965
ISBN 0-913392-77-4

Preface

The NCRP has a long history of involvement with protection against radionuclides that gain entry to the human body. Indeed, the second report issued in 1934 was concerned with radium protection. Subsequently, a number of reports concerned with particular problems of internal emitters led up to the compilation, in 1953, of maximum permissible body burdens and concentrations for a considerable list of radionuclides. This, in turn, was followed in 1959 by a compilation of an even more comprehensive list of radionuclides published as NCRP Report No. 22, *Maximum Permissible Concentrations of Radionuclides in Air and Water for Occupational Exposure*. An addendum was added in 1963. In 1971 the NCRP reviewed, revised, and simplified its basic radiation protection criteria (NCRP Report No. 39), but only general aspects of internal emitter standards were considered. Detailed consideration was the responsibility of specialized internal emitter committees, and their work has resulted in the publication of reports treating various topics related to internal emitters. However, there has been no recent NCRP report broadly reviewing the philosophy and methodology of limiting exposure from internal emitters. This report serves that purpose.

The introduction of a modified system of radiation protection by the International Commission on Radiological Protection (ICRP) in its Publication 26 indicated that important changes were being proposed in the control of internal emitters. ICRP Publication 26 was followed by Publication 30, *Limits for Intakes of Radionuclides by Workers*. The NCRP was then in the midst of its own review and evaluation of radiation protection concepts for internal emitters. However, because potential impact of the new ICRP system was so great, a review of that system was given precedence over the development of other approaches.

In this report, primary concepts relating to protection from internal emitters are reviewed. These primary concepts include dose equivalent, effective absorbed energy and specific effective energy, intake patterns, committed dose equivalent, dose equivalent commitment and population dose, the critical organ concept and committed effective dose equivalent, and stochastic and non-stochastic effects. An evaluation and recommendation is also presented for each of the primary concepts reviewed. The forms of expression of internal emitter standards such as annual limit on intake, derived air concentration and maximum permissible concentration, and derived organ or body burden are considered and evaluated and conclusions stated.

Some attention is devoted to the various deposition, metabolism, and anatomic models which form the basis for an internal dose calculation system. The evolution of current ICRP models is traced, and areas of improvement and limitations on NCRP acceptance of these models are identified. The models evaluated are, reference man, respiratory tract, gastrointestinal tract, bone, radionuclide biokinetics and excretion, and submersion in a radioactive cloud. Excerpts relating to the pertinent models are included in appendices to the report.

Also included in the report is a qualitative statement of present NCRP thinking on control of internal exposure and research needs identified in the preparation of the report.

This report was prepared by Scientific Committee 57 on Internal Emitter Standards. Serving on the Committee were:

J. Newell Stannard, *Chairman*
University of California
San Diego, California

Members

John A. Auxier
Applied Science Laboratory
Oak Ridge, Tennessee

William J. Bair
Batelle Pacific Northwest Laboratory
Richland, Washington

Patricia W. Durbin
University of California
Berkeley, California

Keith J. Eckerman
Oak Ridge National Laboratory
Oak Ridge, Tennessee

Roger O. McClellan
Lovelace Inhalation Toxicology Research Institute
Albuquerque, New Mexico

Paul E. Morrow
University of Rochester
Rochester, New York

Robert A. Schlenker
Argonne National Laboratory
Argonne, Illinois

Roy C. Thompson
Battelle Pacific Northwest Laboratory
Richland, Washington

Consultants

Ian T. Higgins
University of Michigan
Ann Arbor, Michigan

Chester R. Richmond
Oak Ridge National Laboratory
Oak Ridge, Tennessee

NCRP Secretariat **E. Ivan White**
James A. Sphan

The Council wishes to express its appreciation to the Committee members and consultants for the time and effort devoted to the preparation of this report.

Warren K. Sinclair
President, NCRP

Bethesda, Maryland
March 15, 1985

Contents

1. Introduction

The last general report from the National Council on Radiation Protection and Measurements (NCRP) on exposure limits for radionuclides, *Maximum Permissible Body Burden and Maximum Permissible Concentrations of Radionuclides in Air and Water for Occupational Exposure*, was issued in June 1959 (NCRP, 1959) with an addendum published in August 1963. It was essentially conjoint with Publication 2 of the International Commission on Radiological Protection (ICRP, 1959); the relevant ICRP and NCRP committees had the same chairman and several members in common.

When the NCRP reviewed, revised, and simplified its basic radiation protection criteria and published the resulting recommendations in Report No. 39 (NCRP, 1971) only general aspects of radionuclide (internal emitter) standards were considered. Detailed discussion was delegated to specialized internal emitter committees, which since have produced reports on specific radioactive elements (NCRP, 1975b, 1977a, 1977b, 1978, 1979a, 1979b, 1983, 1984a, 1984b), on instrumentation for measuring radiations from radionuclides (NCRP, 1976, 1985a), on special problems of lung dose (NCRP, 1975c), of nuclear medicine (NCRP, 1970, 1982, 1985b), and of other aspects of radionuclide exposure (NCRP, 1975a, 1980b). There has, however, been no NCRP report broadly reviewing the philosophy and methodology of internal emitter standards since 1959. Such a review has been underway within committees and task groups of the NCRP for several years. That review has involved examination of a great deal of new knowledge and new attitudes toward standards, especially the overt consideration of risk and risk/benefit analysis.

The introduction of a modified system of radiation protection by the ICRP in its Publication 26 (ICRP, 1977) presaged important changes in the control of internal emitters. These changes were implemented in ICRP Publication 30, *Limits for Intakes of Radionuclides by Workers*, the first part of which was issued two years later (ICRP, 1979a, 1980b, 1981b). Although NCRP was then in the midst of its own review and evaluation of radiation protection concepts for internal emitters, the appearance of the new ICRP system had such a special impact on internal emitter standards that a review of that system was given precedence.

Current and future work of the NCRP on internal emitter standards may be considered to fall into three phases. Phase I concerns the review and evaluation of past and current concepts for internal emitter protection standards. This report contains the results of Phase I activities. Phase II will be concerned with detailed recommendations for internal emitter exposure control. Phase III will be concerned with numerical limits for specific radionuclides. It is anticipated that Phases II and III will build upon the compendia of biological and physical data contained in ICRP Publication 30 and its supplements (ICRP, 1979a, b, 1980b, 1981a, b, 1982); however, the metabolic models and metabolic constants for all important radionuclides therein will be examined in the light of developing information, and modifications will be recommended when considered necessary.

In addition to these efforts on standards for occupational, and eventually the difficult problem of population exposures, the NCRP is examining environmental transport models and is addressing many specific problems such as disposal of radioactive wastes and the more practical problems of implementing standards.

2. Scope of Report

This report is a necessary, formal update of the positions held and recommendations made in the 1959 NCRP report on internal emitters (NCRP, 1959), including a review of the evolution of internal emitter protection practices in the United States over the past two decades.

The new ICRP methods of calculating radiation doses from internally deposited radionuclides are accepted as a conceptual basis for future standards, but with significant reservations as to specific application. The models employed by ICRP for dosimetry, deposition, transport, and anatomical parameters are adopted as generally superior to those recommended by the NCRP in 1959. However, some deficiencies in these models are identified.

This report is organized into four further sections. Section 3 deals with primary concepts, reviewing their evolution and stating the position of the NCRP on these concepts, particularly as it may coincide with, or differ from, that of the ICRP. Section 4 considers the output of the system—the forms of expression of internal emitter standards, and the manner in which these standards are properly used. Section 5 considers in some detail the deposition, metabolism, and anatomical models which form the basis for an internal dose calculation system. The evolution of current ICRP models is traced, and areas of needed improvement and limitations on NCRP acceptance of these models are identified. Section 6 identifies research needs and Section 7 presents a concise qualitative statement of present NCRP thought on control of internal exposure. Appendices include a consideration of the effect of dosage regimes on the radionuclide content of tissue, information on the central concepts of the new ICRP dose limitation system as they relate to control of internal emitters, the new ICRP methods for calculating doses to tissues from internally deposited radionuclides, and a list of the NCRP task groups contributing to this report.

It is also pertinent to emphasize what this report does not contain. It does not contain a detailed quantitative NCRP system for control of internal exposure; this is being developed as part of a general system encompassing both internal and external irradiation, and will be the subject of a separate report. It also does not contain numerical limits for control of exposure to radionuclides or specific recommendations for evaluation of individual exposures.

3. Primary Concepts

This section considers a number of concepts critical to the formulation of a system of radionuclide exposure standards. Those discussed are new or have undergone significant change in the past two decades. Attention is directed to current ICRP use of these concepts and to the position of the NCRP on such use.

3.1 Dose Equivalent

Previous reports acknowledged the need to modify absorbed dose in rads to account for differences in relative biological effectiveness (RBE) of radiations of different quality (NCRP, 1959; ICRP, 1959). It was stated that, "the rem corresponds to the dose in tissue which results in biological damage equivalent to that produced per rad of x-radiation (of about 200 kV) having a linear energy transfer, LET, to water of 3.5 keV/μ, i.e., rem = RBE × rad" (ICRP, 1959). The product of RBE and rad was included in the "RBE dose," which soon also incorporated factors for the total energy absorbed in the body per disintegration, the ratios (F) of the disintegrations of daughter products, if any, to those of parents and other modifying factors. The expression, $\Sigma EF(\text{RBE})n$, therefore represented the product of the weighted average total energy per disintegration and the appropriate RBE for each radiation type.

Use of the term RBE in both radiobiology and radiation protection was a source of confusion even before the 1959 reports. The awkwardness of the term "RBE dose," and the growing realization that the RBE for different biological endpoints was not the same even for radiations of the same quality, led to designation of the more general term "dose equivalent." To avoid confusion with experimentally determined ratios of radiation effectiveness, the term RBE, as used in radiation protection, was replaced by "quality factor," originally designed as "*QF*" and later by "*Q*." The ICRP and the International Commission on Radiological Units (ICRU, 1968) agreed that

$$(DE) = D(QF)(DF)\cdots \qquad (3\text{-}1)$$

where *DE* is dose equivalent in rems, *D* is absorbed dose in rads, *QF* is the quality factor, and DF is a dose distribution factor; the formulation implies a possible future need for other modifying factors (ICRU, 1963).

The NCRP promptly adopted QF in place of RBE, and also its subsequent contraction to Q, and tacitly adopted "DE" instead of "RBE dose," in Report No. 39 (NCRP, 1971).

A third modifying factor, n, was used in the 1959 ICRP/NCRP reports to calculate the biologically effective energy for bone-seeking radionuclides. That factor took account of experimentally observed differences in the induction of bone tumors by several bone-seeking radionuclides. The differences in bone tumor induction were considered to be due to differences in distribution of the elements within bone. Earlier called the "non-uniform distribution factor," and later the "relative hazard factor," or just "the distribution factor," n is a specific example of a dose distribution factor (DF) as used in the general equation (3-1) above. In Report No. 39, the NCRP accepted the rationale for such a factor but reserved judgment as to its general usefulness (NCRP, 1971).

Another issue relating to dose distribution is the so-called "hot spot" or "hot particle" problem. What are the radiobiological consequences of radionuclide concentration in relatively small volumes of tissue within an organ as compared to diffuse distribution of the same amount of radionuclide throughout the whole organ? Stimulated by the contention that plutonium particles deposited in the lung might induce more lung tumors than predicted on the basis of their average radiation dose to the lung (Tamplin and Cochran, 1974), this issue was analyzed in detail by a number of organizations, including the U.S. National Academy of Sciences (NAS, 1976), the U.S. Atomic Energy Commission (USAEC, 1974), the German Ministry of the Interior (GMI, 1978), the British National Radiological Protection Board (Dolphin, 1974, 1975), and the ICRP (1977, 1979a) and the NCRP (1975c). The common conclusion was that concentration of radiation dose was more apt to decrease than to increase carcinogenic risk, since concentrated sources would irradiate fewer cells and kill a higher proportion of those irradiated. An exception might be the case where the region of localization of dose coincided with a radiosensitive region, as may be the case with surface-seeking radionuclides deposited in bone. The latter case was managed with the n factor in 1959 and is avoided in ICRP Publication 30 by calculation of dose to bone surfaces.

The dose equivalent as now defined by ICRP (ICRP, 1977; ICRU, 1973) has been modified symbolically to:

$$H = DQN \tag{3-2}$$

where H is dose equivalent, D is the absorbed dose, Q is the quality factor and N is the product of all other modifying factors. The units of H are rems if D is in rads, sieverts if D is in grays. ICRP does not use any modifying factors at present. The factor n, discussed above, is not required in the ICRP system because the ICRP accommodates nonuniformity of dose to bone by calculating dose to specific anatomical structures. The quality factor and all the other modifying factors are intended only for use in radiation protection.

While ICRP makes no current use of other modifying factors, N (Eq. 3-2), NCRP recently suggested introduction of "dose rate effectiveness factors" (DREF), (NCRP, 1980a). These are factors by which the total absorbed dose would be reduced to take account of the reduced biological effectiveness of low-LET radiation at low doses and low dose rates. A range of DREF values from 2 to 10 was suggested when absorbed dose was less than 20 rads and/or the dose rate was 5 rads per year or less.

3.1.1 *NCRP Evaluation and Recommendation*

The NCRP suggests that the expression for dose equivalent and the ICRU symbolism of Eq. (3-2) be used in current USA practices. Whether, or to what extent, such a formulation would be used in a revised system of radiation exposure standards will be addressed in the future.

In agreement with its earlier position (NCRP, 1975c), and in the absence of experimental evidence to the contrary, NCRP considers that there is no enhanced carcinogenic risk from "hot particles," and recommends that, for purposes of radiation protection, radiation dose should be averaged over the target tissue, which will usually be an entire organ but may in particular instances consist of a specified volume of radiosensitive cells.

3.2 Effective Absorbed Energy and Specific Effective Energy

In the 1959 ICRP/NCRP reports, the "effective absorbed energy" from internally deposited radionuclides was calculated for the body as a whole, and for significantly irradiated organs on the basis of the radionuclide contents of those organs. It was recognized that such organ doses in some cases were underestimates, to the extent that a part of the dose to the organ was contributed by photons originating from radionuclides present in surrounding tissues or organs. The increasing sophistication of computer applications has made it possible, gradually, to replace "effective absorbed energy" used in the 1959 system with calculations of "specific effective energy" (SEE) which includes "specific absorbed fractions" of photon energy. Calculations of SEE for target organs, that are based on more complete knowledge of the disintegration schemes of radionuclides than were available in 1959, include the energy contributed by photons originating in all significant source organs, as well as the photons and particulate radiations originating from radionuclides deposited in the target organs themselves. This approach to photon dosimetry and the SEE calculations is due largely to Snyder and coworkers (Snyder *et al.*, 1969; Snyder, 1970), who developed them, partly for uses in nuclear medicine by the Medical Internal Radiation Dose (MIRD) Committee of the Society of Nuclear Medicine (MIRD, 1968, 1969a, b, 1970, 1971) and also for use in other U.S. programs e.g., the Reactor Safety Study (USNRC, 1975), the Plowshare Program, numerous hearings, environmental impact statements and planning by regulatory agencies—as well as for the ICRP.

SEE is defined in Section 4.5 of ICRP Publication 30 (ICRP, 1979a), which is reproduced in Appendix B-1 of this report. Values of SEE for various target and source organs are tabulated, for the radionuclides considered, in the Supplements to ICRP Publication 30 (ICRP, 1979b, 1981a, b, 1982). These values of SEE result from the summation over all radiations produced per transformation of a given radionuclide in the several source organs, multiplied by the fractional energy absorption in the target organ and the quality factor for the radiation type, and divided by the mass of the target organ as taken from ICRP Report 23 on Reference Man (ICRP, 1975).

3.2.1 *NCRP Evaluation and Recommendation*

The NCRP considers the SEE calculations as presented in the cited ICRP reports and predecessor documents a general advance over the previous formulation and recommends their use as shown in Appendix B-1.

3.3 Intake Patterns

The 1959 ICRP/NCRP reports based their calculation of derived limits, designated as Maximum Permissible Concentration (MPC) for radionuclides in air (MPC_a) and water (MPC_w), on the assumption of constant intake for 50 years during the 8-hour day of a normal work week (40 hours per week) or continuous intake (168 hours per week). The MPC's of radionuclides in air or water were calculated so that the limiting radiation dose rate to the critical organ or, in some cases the whole body, would not be exceeded at any time during a 50-year working lifetime. In the case of short-lived radionuclides, or those that were rapidly eliminated from the body (short effective half-life), the limiting dose rate might be attained within a few days or weeks. Radionuclides with long half-lives and prolonged retention times (long effective half-life) were permitted to approach the limiting dose rate much more slowly. In some cases the maximum permissible organ burden (i.e., the organ burden which would deliver the limiting dose rate) would be reached only at the end of the assumed 50 years of continuous occupational exposure.

No mechanism was given at that time for calculating the consequences of, or determining limits for, a single intake or a few closely paced intakes. Committee 4 of ICRP addressed single intakes in Publication 10 (ICRP, 1968a). Radiation dose was calculated from the time integral of radionuclide deposition, expressed in μCi- days resulting from the deposition of 1 μCi, using suitable constants to convert from μCi-days to absorbed radiation dose. For occupational exposures the integration period chosen was 50 years.

ICRP Publication 10A (ICRP, 1971) addressed the kinetics of several intakes in a limited interval along with the translocation of radionuclides to other organs resulting from initial contamination of the lungs or a wound. It was shown that widely spaced intakes (i.e., separated by three or four effective half-lives) needed no special formulation and could be treated as individual incidents. It was further shown that as long as the irregularly spaced intakes are not too asymmetrically distributed in size and time, the rate of intake can be considered constant.

Because the subject of dose equivalent from fairly closely spaced intakes has been a source of some confusion and concern in practical health physics, a further consideration and illustrative example is provided in Appendix A.

The ICRP no longer recommends that internal emitter exposure limits be based on continuous intake but rather on the time integrals of dose following single intake. The trend in the United States during recent years has been to use the time integral of dose and the concept of "committed dose equivalent" which is considered below. As illustrated in Appendix A, the pattern of intake has little effect on the long-term accumulation of dose equivalent, and therefore is not fundamental to the derivation of radiation protection limits. However, the intake pattern does influence short-term dose and thus may be important in the assessment of dose from specific exposures.

3.4 Committed Dose Equivalent

The sum of the products of the time integrals of residence of a radionuclide in all source organs (in μCi-days or Bq-days) and the dose equivalent rates per unit deposition of a radionuclide in a target organ (in rem per μCi-day or sieverts per Bq-day) is the dose equivalent to that target organ over the time interval. When the time interval is 50 years as for occupational exposure, this term is defined by ICRP as "committed dose equivalent (H_{50})," for which the formulation taken from ICRP Publication 26 p. 6 (ICRP, 1977), is:

$$H_{50} = \int_{t_0}^{t_0+50y} \dot{H}(t)\, dt, \qquad (3\text{-}3)$$

where $\dot{H}(t)$ is the relevant dose-equivalent rate and t_0 is the time of intake. A more detailed consideration of H_{50}, as used in ICRP Publication 30, is reproduced in Appendix B-2.

The supplements to ICRP Publication 30 (Parts 1, 2, and 3) give abbreviated decay schemes, specific effective energies, the numbers of transformations in source organs during 50 years following ingestion or inhalation of unit activity, and committed dose equivalent (H_{50}) in the organs of interest for specific radionuclides (ICRP, 1979b, 1981a, 1982).

3.4.1 *Previous Applications (Particularly in the United States)*

The concept of dose commitment[1] did not arise *de novo* in ICRP Publications 26 and 30. It was introduced as early as 1958 within the Secretariat of the United Nations Scientific Committee on the Effects of Atomic Radiation (UNSCEAR), primarily to deal with the consequences of the radionuclides in fallout from nuclear weapons tests. Its use in UNSCEAR documents has continued (UNSCEAR, 1977). It was used in 1966 in ICRP Publication 9 (ICRP, 1966a) and subsequently, as a way to deal with the "legacy of a single intake."

In the United States there have gradually evolved tables of dose

[1] See Glossary (Appendix E) and further discussion for distinctions between this term, committed dose equivalent, etc.

equivalents and related derived quantities which utilize the concept of a dose commitment legacy (i.e., committed dose). These differ in significant detail from each other and from the current ICRP usage. Since the reports containing these tables have been used extensively in the U.S. and are still being used, it is appropriate to review them in some detail in order to avoid confusion, particularly unrecognized confusion, in current practices.

The reports discussed below are a sampling of some of the most often used compilations of calculated radiation doses from radionuclides. They reflect the information available at the time, and the special needs for which they were prepared. Unfortunately, no single compilation replaces them all, not even the newest ICRP formulations.

(1) The Medical Internal Radiation Dose (MIRD) Committee of the Society of Nuclear Medicine has published a series of pamphlets, (MIRD, 1968, 1969a, 1969b, 1970, 1971) primarily for use in nuclear medicine in which methods were developed for calculating "absorbed fractions" of photon energy in a tissue from a radionuclide source located in any body organ. The MIRD pamphlets are limited to calculations for photons, beta particles, electron capture and other low-LET emissions; alpha emitters are not discussed because they are not used in nuclear medicine diagnostic procedures. The MIRD formulations are especially useful in that they show the decay schemes and dose information for each radionuclide in a form that can be readily applied to dose estimation in nuclear medicine.

(2) Building on the MIRD reports, Snyder and colleagues (1974, 1975) tabulated in a two-part document the dose equivalent in 24 target organs due to 1 μCi-day deposited in a source organ for 160 radionuclides ($Z \leq 68$). The dose was averaged over the target organ; it applied to adults only, since it was based on a 70-kg phantom. The 22 source organs included cortical bone, cancellous bone, and red and yellow marrow. In order to use Spiers' (1974) method for calculating beta-ray dose to endosteal cells and marrow in the skeleton, all the beta-gamma emitting radionuclides were assumed to be uniformly distributed in bone. The 24 target organs included endosteal cells within 10 μm of bone surfaces, red and yellow marrow, and total bone (average dose to 5000 g of mineralized tissue).

(3) Killough and coworkers (1978) carried Snyder's work forward and prepared estimates for the Nuclear Regulatory Commission of committed dose equivalent, H_{50}, following ingestion or inhalation [three particle sizes, 0.3, 1.0, and 5.0 μm activity median aerodynamic diameter (AMAD)] of 1 μCi of 68 radionuclides of mass number less than 150. With two exceptions (muscle and skin were not considered

separately from the whole body), the 22 target organs were those for which dose equivalent per μCi-day had been tabulated by Snyder, *et al.* (1974, 1975). Dose equivalent in total bone and the three skeletal target tissues was included, calculated as in (2) above.

Deposition in and transport into the body from the respiratory tract was calculated using the model of the Task Group on Lung Dynamics (ICRP, 1966b); and the model of Eve (1966) was used to calculate dose equivalent to the tissues of the gastrointestinal tract. The general kinetic models (and their specific parameters for individual elements) were those developed by S. R. Bernard for the Reactor Safety Study (USNRC, 1975).

(4) Continuing the work of Killough *et al.* (1978), Dunning and coworkers (1979) compiled committed dose equivalent, H_{50}, for 78 radionuclides of mass number greater than 200 (of importance in light-water reactor fuel cycles). Beta-gamma emitters were treated in the same manner as in Killough *et al.* (1978). Alternative tables were given for two different alpha quality factors, $Q = 10$ and $Q = 20$. Alpha emitters in bone were classified as "surface" or "volume" seekers, and committed dose equivalents to endosteal cells and red marrow were calculated using the method of Thorne (1977).

(5) A different set of calculations was prepared by Hoenes and Soldat (1977). Here, age-specific radiation dose commitment factors were given for infants, children, adolescents and adults, due to a continuous intake of radionuclides for one year by inhalation or ingestion. The results were given in units of millirem per 50 years per pCi intake during one year. This compendium was based primarily on the 1959 ICRP/NCRP reports as updated by ICRP Publications 6 (ICRP, 1964) and 10 (ICRP, 1968a) plus physiological and anatomical data from ICRP Publication 23 on Reference Man (ICRP, 1975). Except for the noble gases, the lung model was that of the 1959 reports. Effective energies were based on the 1959 system, where the energy considered is that absorbed in the target organ from nuclear transformations only in that organ. The importance of this compendium is its introduction of age as a factor and, thus, its applicability to a general population. For this reason it continues in general use.

Differences in application of basic principles make the above reports idiosyncratic in details of formulation and, therefore, in ultimate calculated values. This must be understood before results obtained by their use can be compared or combined. Unfortunately, such understanding is not often reflected in the myriad of calculations associated with radiation dose assessment, environmental impact statements, facilities planning, etc. It is unlikely that anyone has been harmed by the disparate numbers involved in these tabulations and formulations. Nevertheless, use of a more uniform system of dose calculation is to be preferred.

3.4.2 *Problems in the Application of Committed Dose Equivalent (H_{50})*

The committed dose equivalent (H_{50}) has been the subject of some confusion and controversy; e.g., Healy (1981). Much of this difficulty would be eliminated if its use was restricted to the calculation of derived radionuclide limits for the purpose of limiting radiation risk, as in ICRP Publication 30. The convenience of committed dose equivalent (H_{50}) as an integrator of complex dose relationships, and the ready availability of calculated values in various publications (see Section 3.4.1) has led to its use in the evaluation of individual and collective risk from radiation exposure, often without regard to the appropriateness of such application.

It can be shown that the committed dose equivalent (H_{50}) to an organ over a 50-year period following one intake, I, of any radionuclide is numerically equal to the dose equivalent rate attained after 50 years of continuous intake at an intake rate, I/year. Thus, derived limits based on committed dose equivalent (H_{50}) would not be numerically different from derived limits which were based on attained dose rate after 50 years of continuous uniform intake. Both approaches result in a somewhat more conservative treatment of long-lived radionuclides that are retained for long periods (long effective half-life) than of short-lived radionuclides or those rapidly lost from the body (short effective half-life). The latter will quickly build up in the organ to the dose equivalent limit, whereas the former will approach this limit more slowly, and it will be necessary to hold their level below this limit until equilibrium is attained, or until the 50-year occupational period has elapsed. This can amount, in the extreme case, to as much as a two-fold difference in accumulated organ dose over the 50-year period; this difference in accumulated dose will, of course, decrease as the person lives beyond the 50-year exposure period. It should also be noted that this "safety factor" for long-retained radionuclides is maximally realized only if exposure occurs at a constant rate.

The principal difference contributed by the committed dose equivalent (H_{50}) of ICRP Publication 30 is that it allows the presumed future consequences of exposure to be included in the single annual committed dose equivalent value, that can be compared with an annual limit on radionuclide intake. The 1959 ICRP/NCRP system, on the other hand, was based upon the organ dose equivalent rate after 50 years, and on the "permissible organ burden" which corresponded to that dose rate.

From the regulatory viewpoint of determining the employer's compliance with standards, the approach based on committed dose equivalent (H_{50}) has advantages. Not only are the presumed future consequences of the exposure explicit in the annual number but also, as we shall see in Section 3.6, it is possible to add presumed risks for different organs and for external exposure. These advantages do not necessarily carry over into the areas of radiation protection responsibility to the employee, where one is concerned with the evaluation of exposure consequences to individuals. In this area actual organ dose equivalents—past, present and future—not committed dose calculated for some arbitrary interval are of primary concern. In this area it is important that information not be lost by "lumping" or "weighting" procedures, or by automatic use of Reference-Man parameters, and that primary attention be focused on actual measured quantities in the individual involved. In the case of internally deposited radionuclides, these measured quantities are primarily environmental media concentrations, secondarily, organ burdens or body burdens and hardly ever intakes. The variability and consequent uncertainty of individual body or organ burdens, even if the intake is known well, reinforce the need to use measures other than committed dose equivalent for the assessment of individual employee exposures.

3.4.3 *NCRP Evaluation and Recommendation*

The NCRP considers the committed dose equivalent (H_{50}) a useful quantity for radiation protection planning and for the demonstration of compliance with those plans, and recommends its use for such purposes. However, committed dose equivalent (H_{50}), involving as it does an extrapolation of cumulative dose to 50 years in the future, does not constitute an appropriate or sufficient basis in itself for the evaluation of radiation exposure consequences *in individuals.* Such evaluations should be based on estimates of actual absorbed dose and the period of exposure appropriate to the individual case. Such estimates of absorbed dose are appropriately compared to annual dose equivalent limits or derived organ burden limits, but this does not imply blanket application of either of these to replace judgement of each exposure individually.

3.5 Dose Equivalent Commitment and Population Dose

ICRP Publication 26 distinguishes between the specific concept of "committed dose equivalent," considered in the previous section, and a more general concept, "dose equivalent commitment," which it defines as the infinite-time integral of the *per caput* dose equivalent rate in a given organ or tissue for a specified population (ICRP, 1977). This more general concept is applicable primarily to considerations involving the exposure of large populations, and necessarily involves assumptions regarding the recycling of radionuclides in the environment, such as have been discussed by UNSCEAR (1977, 1982). This concept is not used in ICRP Publication 30, which is concerned only with limits for workers. The application of this concept will be examined in connection with the development of population exposure standards.

3.6 Critical Organ Concept and Committed Effective Dose Equivalent (Including Organ Weighting Factors)

The 1959 ICRP/NCRP reports employed different dose limits for total body and for several categories of organs and tissues. Total body or organ radionuclide burdens corresponding to these dose limits were calculated, as were the radionuclide concentrations in inhaled air or ingested water that would result in these limiting body and organ concentrations. Such calculations were made for all organs for which there were data, but results were listed only for those receiving the higher doses. The organ for which calculations resulted in the lowest maximum permissible concentrations in air or in water was placed in bold type in the listings and became the limiting, or "critical" organ. Occasionally, relative radiosensitivity or the importance of the organ to life functions modified the choice, especially if two organs led to nearly the same maximum permissible concentrations. Thus, in the 1959 ICRP/NCRP system, one organ, in effect, controlled the choice of allowable concentration in air or water, except when the whole body was the "critical organ." Errors introduced by this simplified approach were considered acceptable in view of the many other uncertainties involved in the calculations.

ICRP Publication 26 prescribes a single dose equivalent limit for the total body and applies this limit to internally deposited radionuclides by summing the doses to various organs, each organ dose being weighted according to the proportion of risk (for cancer and hereditary defects) from uniform exposure of the whole body that is attributable to that organ. This weighted sum of organ committed dose equivalents (H_{50}) has been termed the "committed effective dose equivalent" (ICRP, 1984), although this terminology was not introduced in ICRP Publications 26 or 30. The ICRP formulation and its recommended values for organ dose weighting factors are reproduced in Appendix B-3 of this report (ICRP, 1979a).

A term with the same basic meaning but not tied to a specific time span was defined by ICRP (1978) as the effective dose equivalent (H_E).

In a system where exposure to internally deposited radionuclides is controlled by the same standard, e.g., risk, that is applied to external exposure, it becomes possible to sum external and internal dose and thus limit total exposure as in ICRP Publications 26 and 30.

3.6.1 *NCRP Evaluation and Recommendation*

The NCRP considers the concept of effective dose equivalent and committed effective dose equivalent an advance over the previous use of a single critical organ for calculating derived limits. The effective dose equivalent should not, however, be substituted for a consideration of organ or tissue doses in evaluation of exposure consequences in individuals. The above does not imply unconditional acceptance of all the details of the ICRP "effective committed dose equivalent system." These specifics will be addressed in future NCRP reports.

3.7 Stochastic Effects and Non-Stochastic Effects

ICRP Publication 26 distinguishes between two types of radiation effects: stochastic effects, including malignant and heredity disease, for which the probability of the effect occurring is considered to be a function of dose without threshold; and non-stochastic effects, such as cataract, for which a threshold dose appears to exist. Different dose equivalent limits are prescribed for these two types of effects; for example the "stochastic limit" of 0.05 Sv (5 rem) is applied to the committed effective dose equivalent and the "non-stochastic limit" of 0.5 Sv (50 rem) is applied to individual organs or tissues (except 0.15 Sv for the lens of the eye) (see Appendix B-3). Although the non-stochastic limit is normally higher than the stochastic limit, it is, frequently the determining factor in establishing the annual limit on intake of radionuclides for organs or tissues of low sensitivity to stochastic effects or usually high susceptibility to non-stochastic effects. This is true for most radionuclides that concentrate in bone, thyroid, and other low-cancer-risk organs, including the parts of the gastrointestinal tract that receive the greatest radiation doses after ingestion of poorly absorbed radionuclides. Thus, for these radionuclides the ICRP still uses what amounts to a critical organ approach.

3.7.1 *NCRP Evaluation and Recommendation*

The NCRP recognizes the practical need for an overriding or capping limit to prevent excessive exposure of individual organs of low susceptibility to stochastic effects that might occur in a system based entirely on committed effective dose equivalent. The development and application of such overriding or capping limits is being readied for presentation in a future NCRP report.

4. Forms for Expression of Derived Limits

The primary dose equivalent limits for control of radiation exposure have changed less over the years that are pertinent to this discussion than have the ways of expressing these limits in forms adapted to the practical control of exposure the workplace. We are concerned here with changes that have been proposed, or that may be desirable, in the methods of applying the primary limits to derivation of secondary standards.

In the 1959 ICRP/NCRP reports, the derived limit for practical control of radionuclide exposure was the maximum permissible concentration (MPC). Concentrations of a radionuclide in air $(MPC)_a$, or in water $(MPC)_w$, were calculated in a manner such that continuous intake at those levels during 50 years of occupational exposure would just result in exposure of the critical organ to the primary dose equivalent limit. The quantity of radionuclide required to produce the maximum permissible dose rate in any organ was termed the "maximum permissible organ burden"; the quantity of radionuclide in the whole body associated with a maximum permissible organ burden was adopted in practice and was termed the "maximum permissible body burden". The calculation of these derived limits was based on the concepts discussed in the preceding section.

Difficulties in the calculation and application of these derived limits led to the development of alternative expressions. As embodied in ICRP Publication 30, the changes include:

1. Adoption of the Annual Limit on Intake (ALI) as the principal vehicle of radionuclide exposure control.
2. Substitution of the term, Derived Air Concentration (DAC), for the term $(MPC)_a$.
3. Elimination of any statement of $(MPC)_w$.
4. Elimination of any statement of maximum permissible organ burden and maximum permissible body burden.

The NCRP position of these changes will be discussed in this section.

4.1 Annual Limit on Intake (ALI)

Although not identified as such in the 1959 ICRP/NCRP reports, the ALI was, in fact, a necessary intermediate in the calculation of

MPCs. The ICRP has chosen to focus attention on this intake limit, which it considers a "secondary limit," rather than on "derived limits" such as MPCs, for several reasons: (1) The ALI is a more fundamental expression from which derived limits for concentration in any desired environmental medium can be calculated; the derived limits (MPCs) are, on the other hand, limited in their application to a specific medium, e.g., air, water. (2) By diverting attention from MPCs, the ICRP hoped to avoid their widespread misapplication, as will be discussed in the following section. (3) An annual limit on intake became feasible when the ICRP abandoned limits for shorter periods (ICRP, 1977).

4.1.1 *NCRP Evaluation and Recommendation*

While NCRP finds the concept of the ALI valid, and concurs in the arguments presented in the preceding paragraph, it is nevertheless concerned that the ALI may be subject to at least as much misuse and abuse as have been the MPCs. This possibility stems from the fact that rates of intake of radionuclides are often difficult to measure and, in most cases, must be estimated from environmental data and length of exposure. Potential sources of radionuclides are numerous and intake is apt to be accidental, infrequent, and idiosyncratic and the effects depend significantly on chemical form; thus, except for planning purposes, a limit on total radionuclide intake is of little direct practical utility. It is useful as a basis for the calculation of derived limits for radionuclide concentration in exposure media and in exposed persons—limits such as nuclide concentrations in various media and organ or body burdens, which are susceptible to some degree of direct measurement and control.

The NCRP, therefore, views the ALI as a useful basis for the derivation of more directly applicable operational and environmental standards such as the DAC, but considers it impractical as a direct measure of compliance.

4.2 Derived Air Concentration (DAC) and Maximum Permissible Concentrations

A principal difficulty in the regulatory application of MPCs has been the popular misconception that these were limits that could not be exceeded without resulting in overexposure, and that serious health problems would ensue from such overexposure. Exposure over the limits, of course, occurred only when the quarterly-average radionuclide concentration exceeded the MPC, but this point was difficult to reconcile with the "maximum permissible" terminology.

ICRP Publication 30 has retained a quantity essentially identical with the old $(MPC)_a$, but has renamed this quantity the "derived air concentration" (DAC). The new terminology avoids the connotation of "maximum" or "permissible" and denominates the concentration which, if breathed continuously during working hours, would give an intake equal to the ALI. With the working year taken as 2,000 hours, the physical activity as "light", i.e., that identified with a breathing rate of $0.02 m^3$ of air per minute, and 60 as the conversion factor from minutes to hours:

$$\mathrm{DAC} = \frac{\mathrm{ALI}}{(2{,}000 \times 60 \times 0.02)}$$

$$= \frac{\mathrm{ALI}}{2.4 \times 10^3} \, (\mu\mathrm{Ci/m^3}) \; (3.7 \times 10^4 \mathrm{Bq/m^3}) \, . \quad (3\text{–}4)$$

4.2.1 *NCRP Evaluation and Recommendation*

The NCRP considers the new terminology an improvement, and concurs in the need for such a derived limit. Inhalation of contaminated air is probably the most common route of occupational radionuclide exposure and the monitoring of radionuclide concentrations in air is an important, and often the only practical measure of the effectiveness of control.

The NCRP also agrees that there is no need for a derived water concentration for workers since ingestion of radionuclides via water is seldom encountered occupationally. Even for the general population water is only one of many potential routes of radionuclide ingestion all of which should be considered in assessing exposure. Derived limits for any environmental medium can, of course, be calculated from the ALI, with appropriate environmental pathway assumptions, where the situation demands such limits.

4.3 Derived Organ or Body Burdens

The statement of maximum permissible organ and body burdens in the 1959 ICRP/NCRP reports, and in much subsequent literature, has been subjected to the same misinterpretations of "maximum" and "permissible" as has been the case with MPCs. It is clear that depositions of radionuclides of short effective half-times can exceed maximum permissible organ or body burden limits for short periods without exceeding quarterly or annual dose equivalent limits. As in the case of MPCs, use of terms such as "derived" organ or body burden might help to prevent this misinterpretation.

In the context of the ICRP Publication 30 system of radionuclide exposure limits, derived organ or body burdens might take one of two forms. They might relate to the organ or body burden corresponding to the dose equivalent limit, and thus be similar to the old maximum permissible organ or body burdens; they might relate to the ALI (i.e., to a committed dose equivalent limit). Perhaps because of this confusing option, the ICRP has chosen to list no burden limits, although the numerical data required for the derivation of such limits are contained in the supplements to ICRP Publication 30 (ICRP, 1979b, 1981a, 1982) and the computer codes associated therewith.

4.3.1 *NCRP Evaluation and Recommendation*

The NCRP considers that derived organ and body burdens are essential tools of the health physicist, and probably the most relevant guide to the evaluation of individual exposures and their consequences. The absence of such derived burdens from ICRP Publication 30 leaves only the ALI as a "sanctioned" basis for such evaluation, which, as discussed in Section 4.1.1, can lead to many problems. Therefore, it is planned that forthcoming NCRP reports on intake of specific radionuclides will include a statement of the organ burdens (or total body burden) associated with the controlling dose equivalent limit.[2]

[2] It should be noted that bioassay procedures continue to be required in both American and British regulatory codes and that the results of these procedures tie conveniently to organ or body burden.

5. Models for Calculation of Limits

The calculation of exposure limits for hundreds of different radionuclides, the biological behavior of which is often poorly understood, is an exercise requiring many assumptions. These assumptions are incorporated into mathematical models of two general types: those describing the exposed individual and the component organ systems of that individual, and those describing the behavior of specific radionuclides within the individual.

A set of simple models was introduced in the 1959 ICRP/NCRP reports to meet this need for generalization. The authors of those reports recognized fully the degrees of simplification introduced, and surrounded their recommendations with suitable caveats. In the intervening years many modifications have been suggested; in particular, much new information has accumulated on the distribution and retention of radionuclides in the body. The purpose of this Section is to set forth and explain the NCRP's views on the more important of these models, and to suggest modifications that may need to be considered in the future. The discussion of these models will proceed from the more general to the more specific and will include, where pertinent, some review of previous methodologies as well as a statement of present position.

5.1 Reference Man

In the 1959 ICRP/NCRP reports several tables described the characteristics of what was then referred to as "Standard Man." As the result of a monumental effort on the part of several members of the Internal Dose Group at Oak Ridge National Laboratory, functioning as a task group for ICRP Committee 2, a greatly expanded and reorganized compilation of data on "Reference Man"[3] was published in 1975 (ICRP, 1975). Projects to extend these data to nonadults, to the elderly, to people whose body sizes, dietary habits, and environ-

[3] This was a generic term and the characteristics of "Standard Woman" are included.

mental conditions differ from those found in North America and Northern Europe, and other obvious expansions are currently in progress.

5.1.1 *NCRP Evaluation and Recommendation*

The NCRP recommends the use of Reference Man data whenever generalized values are required for such parameters as weights, surface areas, anatomical dimensions, gross and elemental composition of organs and tissues, and physiological data for organs and tissues. As is the case for all models, values should be modified if the conditions of the exposure and/or the characteristics of the individual or the population concerned are known to deviate substantially from those of Reference Man.

5.2 Respiratory Tract Model

The 1959 ICRP/NCRP reports introduced a simple lung model based largely on limited data from experiments in small animals. This model assumed that 25 percent of inhaled particles were immediately exhaled, 50 percent were deposited in the upper respiratory passages and subsequently cleared via the mucociliary mechanism, and the remaining 25 percent were deposited in the lower respiratory tract. This latter fraction (25 percent of inhaled material), if "readily soluble" (in water at pH 7), was assumed to be quickly and completely absorbed into the blood stream. If "insoluble", one-half (12.5 percent of the initial intake) was considered to be eliminated within twenty-four hours by way of the G.I. tract, and one-half was retained in the lungs and was assumed to be taken up into body fluids with a half-life of 120 days (except for thorium and plutonium, then considered to be retained with a half time of 300 days). No account was taken of the effect of particle size, and chemical form was considered only in terms of "soluble" and "insoluble" materials.

Perhaps the greatest deficiencies in the original lung model were its inability to account, (1) for the slow dissolution of many "insoluble" compounds and their subsequent translocation in the body, (2) for the translocation of some highly insoluble particles into the lymphatic tissue of the respiratory tract, and (3) for the effects of particle size and chemical form on clearance rates. Indeed, it has been indicated by former members of the original Committee 2 that the 1959 model could not be and was not always applied in the form stated in Table 10, page 153 of ICRP Publication 2. For example, because of the problems of slow dissolution and translocation stated above for certain long-lived insoluble radionuclides, those preparing the publication did not assume that the 12.5 percent remaining was taken up in the body fluids. Instead, it was assumed that this portion remained in the lung with the much longer effective half-life indicated above or even longer.

Respiratory tract parameters, the effect of particle size on deposition, and material transport from the lung into the body, were studied in depth by a task group of the ICRP, which published a significantly revised model in 1966 (ICRP, 1966b). Some further quantitative modifications were made in ICRP Publication 19 (ICRP, 1972). The "Task Group Lung Model" addresses deposition of particles (as a function of size) in the three major compartments of the adult human respiratory tract—naso-pharyngeal (N-P), tracheobronchial (T-B), and pulmonary (P). Parameters are provided for the clearance of three major classes of inhaled compounds (as defined by their chemical solubility) from the three respiratory tract compartments via the mucociliary pathway, and for the translocation of these radionuclide compounds from the respiratory tract to the lymph nodes (compartment L) and other body organs. This model was a significant advance over the simple 1959 lung model, and was immediately used in calculations concerned with radiation protection.

5.2.1 *Current ICRP Model*

The Task Group Lung Model of 1966, with modified clearance constants as adopted in ICRP Publication 19, was first used officially by the ICRP in Publication 30 (ICRP, 1979a) and amended in 1981 (ICRP, 1981b). Basic principles of the model are illustrated in Figs. 5.1 and 5.2; a detailed description is reproduced in Appendix B-4.

The manner in which this model was used by the ICRP in Publication 30 departed in many details and substantially from the original Task Group recommendations: (1) the T-B, and P and the L regions were combined into a single target tissue with a mass of 1000 grams, and (2) the range of aerosol size considered (Fig. 5.1) was reduced to 0.1 to 10 μm AMAD from the original range of 0.01 to 100 μm AMAD. The ICRP elected to consider T-B, P, and L regions as a single organ for dosimetric purposes because of uncertainties about the precise location of the cells at risk. For example, although the L region often receives the highest dose, malignancies are rarely observed other than the T-B and P regions. The ICRP emphasizes that when special

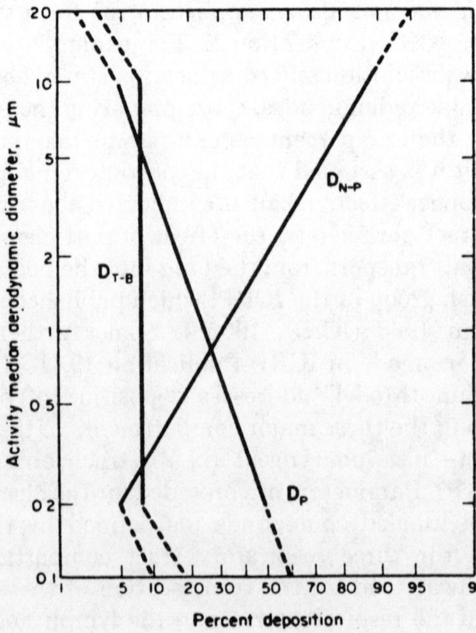

Fig. 5.1 ICRP Publication 30 respiratory tract deposition model (ICRP, 1981b).

Region	Compartment	D: T day	D: F	W: T day	W: F	Y: T day	Y: F
N–P	a	0.01	0.5	0.01	0.1	0.01	0.01
(D_{N-P} = 0.30)	b	0.01	0.5	0.40	0.9	0.40	0.99
T–B	c	0.01	0.95	0.01	0.5	0.01	0.01
(D_{T-B} = 0.08)	d	0.2	0.05	0.2	0.5	0.2	0.99
	e	0.5	0.8	50	0.15	500	0.05
P	f	n.a.	n.a.	1.0	0.4	1.0	0.4
(D_P = 0.25)	g	n.a.	n.a.	50	0.4	500	0.4
	h	0.5	0.2	50	0.05	500	0.15
L	i	0.5	1.0	50	1.0	1 000	0.9
	j	n.a.	n.a.	n.a.	n.a.	∞	0.1

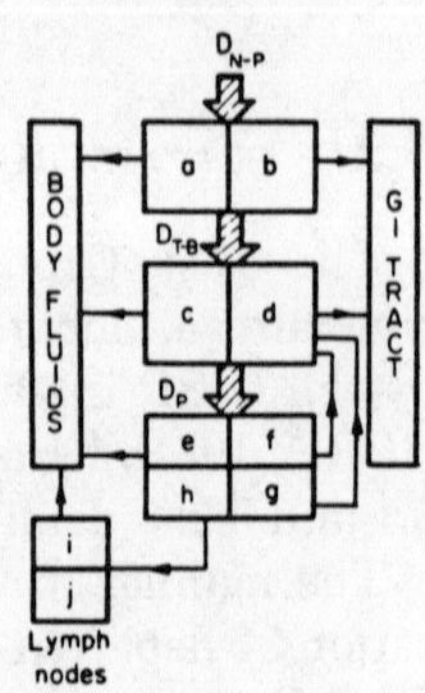

Fig. 5.2 ICRP Publication 30 respiratory tract clearance model (ICRP, 1981b). The symbols N-P, T-B, P and L are defined in Section 5.2.

circumstances indicate localization of the material, or a special sensitivity of cells in a given region of the respiratory tract, the model provides the flexibility required to make such calculations. The narrowed range of particle sizes considered undoubtedly reflects the relative unimportance to dose of particles above 10 μm AMAD and the present difficulty of predicting the behavior of particles smaller than 0.1 μm AMAD. More detail concerning ICRP views on the behavior of radionuclides deposited in the respiratory tract is given in ICRP Publication 31 (ICRP, 1980a).

5.2.2 *NCRP Evaluation and Recommendation*

A task group of NCRP Scientific Committee 57 (see Appendix D for membership) has reviewed the ICRP lung model and other features of the dosimetric approach to the respiratory tract as used in ICRP Publication 30. It has also reviewed other uses of the model since its original publication in 1966. While impressed with the basic features of the model, there is disagreement with some aspects of its application in Publication 30.

The lack of consideration of dose to the nasopharyngeal region is a cause for concern.[4] Although calculated doses to this region may be small in comparison to the doses received by other regions, there is evidence from animal experiments that some radionuclides are appreciably retained in the N-P region and may subsequently cause cancer in this region. The calculation of an average dose to the entire respiratory tract, while expedient for the derivation of exposure limits, is not wholly satisfactory for understanding or predicting effects in the various regions of the lung. The distribution of effects in the lung and accessory respiratory structures is nonuniform, e.g., development of tumors of bronchiolar origin in response to exposure to radon and radon daughter products. The distribution of dose within the various regions of the lung is also nonuniform, e.g., the accumulation of very large concentrations of inhaled insoluble particles in pulmonary lymph nodes. While primary cancer of the lymph nodes following radionuclide inhalation is a rare occurrence in experimental animal studies, the effect of the extensive damage in these lymph nodes on general functioning of the lymphatic system and on immune surveillance of the lung cannot at present be evaluated. Information on the deposition of inhaled particles smaller than 0.2 μm AMAD should soon be available from experiments currently in progress. These and other matters will be discussed in greater detail in an NCRP Report on Respiratory Tract Kinetic Models.

The above deficiencies notwithstanding, the general lung model developed by the ICRP, as adopted in Publication 30, is a major

[4] In ICRP Publication 30 it was considered to be so small that it could be neglected. It was regarded as "unimportant" in the original Task Group report.

advance over the primitive model used in the 1959 ICRP/NCRP reports, and the NCRP recommends use of this new lung model for occupational radiation protection purposes. It should be emphasized that this recommendation, like others in this report, applies only to prospective planning for the control of radiation exposure of workers and not to the retrospective evaluation of individual exposures if and when organ burden can be detected directly. Nor is it directly applicable to the evaluation of general population exposures, because differences atttributable to size of organs, e.g., in the young, the effects of pulmonary disease on both particle deposition and clearance, and other factors in the exposure of populations are not included in the current model.

5.3 Gastrointestinal Tract Model

The 1959 ICRP/NCRP reports gave information on the masses of the gastrointestinal segments and their contents, and the times the contents remained in the stomach, small intestine, upper large intestine, and lower large intestine. In 1966, Eve (1966) and Dolphin and Eve (1966) suggested modifications, which have been generally accepted.

Parameters incorporated in ICRP Publication 30 are shown in Fig. 5.3 taken from the ICRP report. The equations for calculation of committed dose equivalent to sections of the gastrointestinal tract are reproduced in Appendix B-5. Each segment of the gastrointestinal tract is considered to be a single compartment (Fig. 5.3), and transfer to the next compartment is considered to follow first order kinetics. A factor varying between zero and one is applied to account for the degree to which various radiations penetrate the mucus covering the gastrointestinal epithelium. It is taken as unity for photons and β particles, 0.01 for α particles and fission fragments, and zero for recoil atoms.

5.3.1 *NCRP Evaluation and Recommendation*

A Task Group of NCRP Scientific Committee 57 (see Appendix D for membership) has considered the gastrointestinal tract model set out in ICRP Publication 30. The chief reservation relates to the conservatism involved in calculating dose at the interface between intestinal contents and intestinal wall. The proliferating crypt cells,

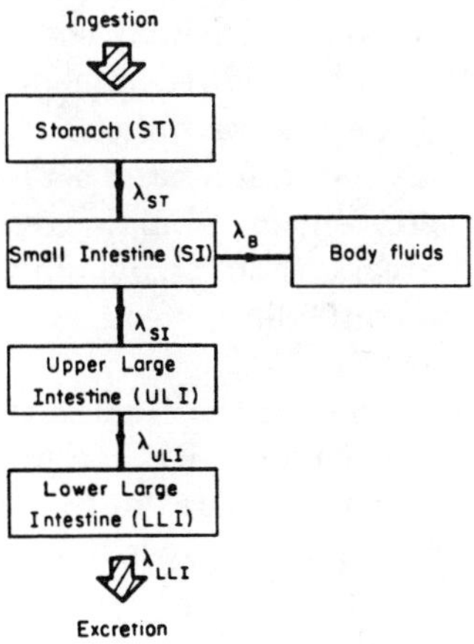

Section of GI tract	Mass of walls* (g)	Mass of contents* (g)	Mean residence time (day)	λ day^{-1}
Stomach (ST)	150	250	1/24	24
Small Intestine (SI)	640	400	4/24	6
Upper Large Intestine (ULI)	210	220	13/24	1.8
Lower Large Intestine (LLI)	160	135	24/24	1

Fig. 5.3 ICRP Publication 30 gastrointestinal tract model (ICRP, 1979a).

which are important to carcinogenesis and cell injury, lie at an approximate depth of 1000 μm within the intestinal wall; thus, for low energy photons or for alpha emitters, the model may substantially overestimate the dose and consequently underestimate allowable limits. This problem is alleviated in the case of alpha emitters by the factor of 0.01 used to account for the larger portion (about 99 percent) of the alpha energy being absorbed in the mucus layer and the smaller portion (about 1 percent) reaching the underlying cells. The NCRP recommends the general use of the gastrointestinal tract model given in ICRP Publication 30 (ICRP, 1979a) with the reservation that critical decisions based on dose estimates for low energy photons, low energy beta particles and alpha particles should be reviewed.

5.4 Bone Models

The 1959 ICRP/NCRP reports calculated limits for bone-seeking radionuclides on the basis of a comparison with the accepted maximum permissible body burden of 0.1 μg of radium-226. In 1965, a task group of ICRP Committees 1 and 2 reviewed the radiosensitivity of the tissues of bone (ICRP, 1968b), and recommended the use of red bone marrow and endosteal bone cells as target tissues for the calculation of maximum permissible levels for hard beta emitters localizing in bone. Alpha emitters were excluded from this recommendation because the calculation of endosteal cell dose from alpha emitters was at that time considered to be too difficult. Instead, comparison to radium was recommended.

A detailed metabolic model for the alkaline earth elements in bone was developed by a task group of ICRP and presented in Publication 20 (ICRP, 1973). The primary features of this model are summarized in Appendix C. Similarly detailed models have not been developed for other bone-seeking elements.

5.4.1 *Current ICRP Model*

The ICRP dosimetric model for bone is based on calculational procedures proposed by Spiers (1974) and by Thorne (1977) and is reproduced in Appendix B-6 of this report, taken from ICRP Publication 30. This model recognizes six categories of radionuclide emissions, as briefly outlined in Table 5.1. These emissions irradiate two

TABLE 5.1—*Principal parameters of ICRP Publication 30 dosimetric model for bone.*

Categories of Radionuclides	Source Organs	Fractional Distribution Within bone	Absorbed Fraction in	
			Bone Surface Cells	Red Bone Marrow
Photon emitters	16 organs and tissues		(Values for various photon energies given in ICRP Publication 23)	
Alpha emitters uniformly distributed in bone	Trabecular bone	0.2	0.025	0.05
	Cortical bone	0.8	0.01	0.0
Alpha emitters deposited on bone surfaces	Trabecular bone	0.5	0.25	0.5
	Cortical bone	0.5	0.25	0.0
Beta emitters uniformly distributed in bone	Trabecular bone	0.2	0.025	0.35
	Cortical bone	0.8	0.015	0.0
Beta emitters on bone surfaces $E_\beta \geq$ 0.2 MeV	Trabecular bone	0.5	0.025	0.5
	Cortical bone	0.5	0.015	0.0
Beta emitters on bone surfaces $E_\beta <$ 0.2 MeV	Trabecular bone	0.5	0.25	0.5
	Cortical bone	0.5	0.25	0.5

targets in bone: the endosteal cells on bone surfaces, considered to occupy a 10 μm thick layer of tissue covering a total bone surface area of 12 m^2, and with a mass of 120 g; and the active red bone marrow within cavities of trabecular bone, with a mass of 1500 g. Photon emissions may arise from any source organ, and the absorbed fractions of photon energy in the bone targets are those given for Reference Man (ICRP, 1975). Alpha and beta emissions arise from either trabecular or cortical bone, where the radionuclides are considered to be deposited according to one of two patterns: uniformly throughout mineralized bone volume, or on bone surfaces. The choice of either the volume- or surface-distribution model is based on the biochemical behavior of each element, except that radionuclides of physical half-life shorter than 15 days are always considered as surface seekers on the basis that they will not have time to penetrate beneath the surface. Thus, all plutonium isotopes are considered to be surface seekers; radium-226 is considered a volume seeker, but radium-224, because of its short half-life, is classified as a surface seeker. The assumed fractional distributions within bone of the various categories are shown in Table 5.1, as are the assumed absorbed fractions in the two target tissues resulting from these distributions.

The parameters of the new ICRP bone dosimetry model, and the basis for their choice, are described in greater detail in Appendix B-6. By these somewhat arbitrary maneuvers the difficult calculation of dose to endosteal cells from alpha emitters has been accomplished by ICRP in a manner satisfactory for the purposes of radiation protection, and a unified system is presented without the awkward necessity of employing different systems for different types of radiation.

5.4.2 *NCRP Evaluation and Recommendation*

A task group of NCRP Scientific Committee 57 (see Appendix D for membership) has reviewed the problems of bone-seeking radionuclides in terms of their metabolism, dosimetry, and biological effects. In the view of the task group, the ICRP metabolic model for the alkaline earths (ICRP, 1973) is no more complex than alternative compartment models (see, for example, Marcus and Becker 1980, or Johnson and Meyers, 1981). As is the case with most models, it may not fit data on specific individuals especially well (Harrison, 1981), but it gives an acceptable fit to population data, particularly with some alteration of the parameters (Schlenker *et al.*, 1982).

The NCRP has some reservations concerning the categorical designation of radionuclides as either surface seekers or volume seekers, when it is well recognized that volume seekers are far from uniformly distributed, and it is unlikely that surface seekers are uniformly distributed on all surfaces, and that they remain there indefinitely. This would seem to be an area in which significant improvement could be made; some efforts in this direction have already been reported (Priest and Hunt, 1979). For the present, however, the lack of consideration of redistribution can be accepted as a factor of conservatism in the model, and is recognized as such by the ICRP (ICRP, 1979a).

The NCRP task group considered in some detail the anatomy of the regions of bone containing the presumed target cells for bone cancer and for leukemia. The flattened cells adjacent to endosteal bone surfaces are thought to be only part of the bone cancer target cell population. It should also include other cells with osteogenic potential that lie deeper within the stroma of the marrow (Kimmel, 1981; Baron *et al.*, 1982). During the process of bone formation, osteoprogenitor cells may be found several cell diameters away from the bone surface—farther than the 10 μm cell thickness assumed by the ICRP. With regard to the presumed target cells for leukemia induction, the task group notes that these marrow stem cells are now thought to be nonuniformly distributed within the marrow; in mice they are concentrated some 120 μm away from the bone surface (Lord, 1975). This has important connotations with respect to the likelihood of leukemia induction by alpha- or low energy beta-emitters deposited within bone or on bone surfaces.

A major concern of the task group in its review of the ICRP system, indeed of any system for bone dosimetry and for the development of radiation protection standards for bone-seeking radionuclides, was the role of the epidemiologic data on radium in man. These data, together with radionuclide toxicity ratios derived from experiments with animals, were the basis for the earlier system of exposure limits for bone-seeking radionuclides. The new ICRP system replaces this approach by calculating absorbed dose to the sensitive tissues in the skeleton as described in the previous section. The new ICRP reports give no indication that epidemiologic experience with radium contributed to the development of the limits for the bone-seeking elements. However, the epidemiologically established maximum permissible body burden for radium, and the radionuclide toxicity ratios derived from animal experiments, have a reliability and cogency at least equivalent to that of the ICRP dose-response relationships. The NCRP considers that both of these bodies of data should be an important component of any scheme for limiting the risk from exposure to bone-seeking radioelements.

Calculation of the dose-equivalent rate associated with the ICRP/

NCRP-1959 limit of 0.1 μg ^{226}Ra, using traditional assumptions regarding distribution and quality factor, leads to a value of about 30 rem/year to a 10 μm-thick endosteal tissue layer lining the medullary cavities and to a value of about 50 rem/year when the Haversian canal endosteal surfaces are included. This calculated dose could be raised or lowered somewhat with other assumptions. This difference was not considered critical at the time. The dual standard basis was continued by both ICRP and NCRP until the issuance of ICRP Publications 26 and 30, when the ICRP dropped direct use of the radium standard.

The NCRP considers that estimation of risk for radium from the accumulated data on man should be possible and would be useful. However it is aware of the many complications involved in this procedure. It has assigned this task to its task group on problems of bone.

5.5 Submersion in a Radioactive Cloud

Submersion in a cloud of radioactive gas results in a mixture of external and internal exposure. The skin and other organs near the surface may be irradiated by external exposure, depending on the penetrating power of the radiation, while internal organs may be exposed by gas inhaled or absorbed through the skin. The respiratory tract will, in addition, be irradiated by the fraction entering the lung; the deeper tissues will be irradiated by the material absorbed into the circulation and distributed to them. Radionuclides for which such a model is particularly important include tritium vapor and the radioactive isotopes of the noble gases (argon, krypton, xenon and radon, and their radioactive daughters when airborne). In an accidental airborne release from an operating reactor, the list could be longer, depending on temperature and conditions of the release. The NCRP is impressed with the careful analyses of Berger (1974), and of Poston and Snyder (1974), as applied in Chapter 8 of ICRP Publication 30. It recommends the use of these analyses for prediction of radiation dose from submersion in a radioactive cloud.

5.6 Radionuclide Biokinetic Models

In the 1959 ICRP/NCRP, reports many simplifying assumptions were made about the distribution and retention kinetics of radionuclides in the body. Values were tabulated for the fraction of each radionuclide reaching blood following either ingestion or inhalation, and for the fraction of that in the blood reaching the compartment of longest retention in the organ of reference. Also tabulated were single effective half-lives describing retention of the radionuclide in several organs and in the whole body. In most cases, the single exponential retention term used for calculation of organ dose was that associated with the slowest rate of release of radionuclide from the organ, which, in some cases, would tend to exaggerate dose. For a few radionuclides, retention described by a power function was included, but only for comparison. The oversimplifications inherent in this system were recognized fully, but it was considered adequate for radiation protection purposes.

In the years since 1959, much new information on the distribution and retention of many radionuclides has become available. The sources include both experiments designed to obtain metabolic and toxicologic data on internal emitters for radiation protection purposes, and information accruing incidentally from the practice of nuclear medicine. These studies frequently consider deposition in, and loss from, more than one compartment of each tissue or organ, and usually assume first order kinetics for the loss from each of these compartments (Jacquez, 1972; Skrable *et al.*, 1974).

In its current calculation of radionuclide limits, the ICRP uses the general biokinetic model shown in Fig. 5.4, consisting of instantaneous uptake followed by first order loss from one or more subcompartments

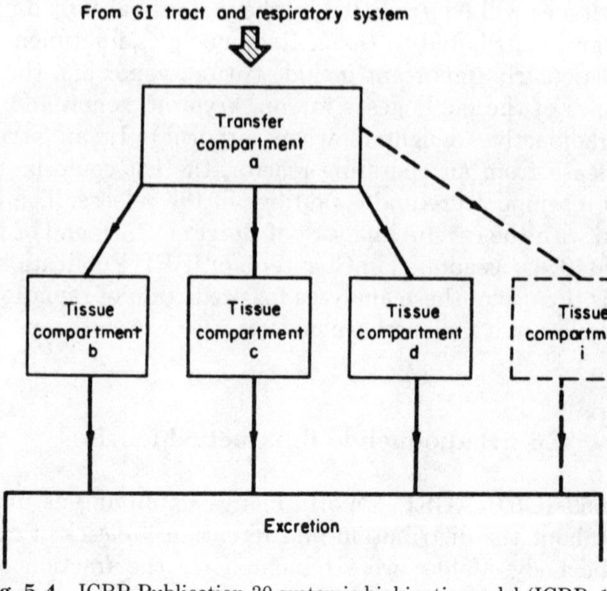

Fig. 5.4 ICRP Publication 30 systemic biokinetic model (ICRP, 1979a).

(ICRP, 1979a). Thus, the metabolism of the radio-element is assumed to follow a system of first order differential equations with constant coefficients and without feedback. The absence of explicit feedback terms, while physiologically inaccurate, is not a source of significant error for radionuclides that are efficiently excreted, or when the biological data used to obtain the model parameters incorporate normal feedback. The ICRP deviates from this general model when better formulations are available; the alkaline earths, for example, are handled by the special model given in ICRP Publication 20 (ICRP, 1973) (see Appendix C).

5.6.1 *NCRP Evaluation and Recommendation*

The current ICRP biokinetic models, as represented generally in Fig. 5.4 and specifically for each element in the Metabolic Data sections of Parts 1–3 of ICRP Publication 30 (ICRP, 1979a, 1980b, 1981b), are superior to, and should be used in preference to, those of the 1959 ICRP/NCRP reports. It must be appreciated, however, that the ICRP model parameters were chosen for their specific applicability to the derivation of radiation exposure limits according to ICRP procedures and they should be used with caution for other purposes.

5.7 Radionuclide Excretion Models

Measurement of radionuclides in excreta is widely used to estimate body content when such content cannot be determined accurately by external measurement. Using the general biokinetic model depicted in Fig. 5.4, total excretion would be the sum of contributions from each of the several tissue compartments and would, presumably, be represented by a sum of exponential terms (Jacquez, 1972; Skrable *et al.*, 1974). In some instances, and over limited intervals, excretion is as well, or better, represented by power functions of time, as for example in the first few years after an intake of plutonium or uranium. Consideration of this aspect of internal exposure evaluation is beyond the scope of the present report, and the issue is raised only to caution against the indiscriminate use of the biokinetic models of ICRP Publication 30 for purposes other than those for which they were developed. Interpretation of excretion data for purposes of body burden estimation should be based on models derived with that application primarily in mind. The models of ICRP Publication 30 were derived for the estimation of organ dose and were not intended to account for excretion.

6. Research Needs

The NCRP has, increasingly, sensed the need to make evident the needs for additional research that are identified in the course of work on NCRP reports. The following list is not intended to be exhaustive and contains some of those research needs identified in the preparation of this report:

(1) Further exploration of the similarities and differences between the effects of a given absorbed dose from deposited radionuclides and external low-LET radiation sources; significance of any differences for the summation of risk from different sources.

(2) Expansion of work to establish dose-response functions in man for a variety of radionuclides, e.g., radioiodine, and other isotopes used for therapy.

(3) Further extension of the parameters for "Reference Man" to a wider range of age and ethnic groups.

(4) Reexamination of the model for lung dynamics with particular reference to the validity of omitting the dose to the nasopharyngeal region and of treating tracheo-bronchial, pulmonary and lymphatic areas as a single organ.

(5) Refinement of solubility classifications of many compounds for the lung model and revision of some of the clearance factors.

(6) Development of lung models for the very young and older members of the population including the influence of pulmonary pathology.

(7) Evaluation of the significance of the difference between assumptions made and the true location of the sensitive cells in the gastrointestinal tract.

(8) Extension of the biokinetic model for alkaline earth elements in bone to other important groups of radioelements.

(9) More definitive localization and identification of the potential target cells for carcinogenesis in bone, in bone marrow and in the respiratory tract with special reference to their geographic relationships to deposition sites of radionuclides.

(10) Reexamination of biokinetic models for many important radio-

nuclides, especially with regard to transfer factors between gut and blood, kidney and urine, etc.

(11) Much more formal and detailed determination and statement of the uncertainties attending the processes and parameters involved in development and application of radiation protection systems for radionuclides in the body.

7. Summary Statement of NCRP Position on Control of Internal Dose (with special reference to ICRP Publications 26 and 30)

The principal objective of an occupational radiation protection system must be the prospective regulation of the workplace. Primary standards are established, which define individual exposure limits, and from which operating standards are derived for control of the workplace environment. The primary standard for internal emitter exposure, as defined in the 1959 ICRP and NCRP reports, was an annual dose equivalent rate to a "critical organ" which was not to be exceeded during 50 years of continuous intake (or, in the special case of bone-seekers, a radionuclide deposition equivalent in biological risk to a retained burden of 0.1 μCi of ^{226}Ra). Derived limits were then calculated specifying concentrations of radionuclides in air or water which, during 50 years of continuous intake, would result in the radiation dose rate (or organ radionuclide burden) set as a limit by the primary standard.

The new system, established by ICRP Publication 26 and applied in ICRP Publication 30, retains radiation dose rate as the primary standard but expresses this dose rate for internal emitters in terms of a committed dose equivalent over a 50-year period rather than as an attained dose equivalent rate after 50 years of continuous intake. This is not a significant change so far as the derivation of operational standards is concerned (see section 3.4.2). The numerical values of these operational standards are controlled primarily by the metabolic and dosimetric parameters used in the calculation. If there were no changes in these parameter values, the operational standards would be the same when calculated by either method. Since many changes in metabolic parameters have been introduced over the period from 1959 to the present and the dosimetry has been modified by including more than one target organ, several source organs and more complete decay schemes in some instances, the derived air concentrations (DAC) of ICRP Publication 30 and its Supplements do differ numerically from their counterparts of 1959. Some are higher, others are lower. However, it is not the system *per se* which generates these differences but rather new data, better analytic techniques and changes in model parameters.

ICRP Publication 26 also modifies the critical organ concept, by taking account of simultaneous irradiation of several radiobiologically important organs, through the mechanism of the effective committed dose equivalent. This is a logical refinement, only partially realized, since many limits are still based on the "nonstochastic" limit, which is effectively a critical organ limit. In terms of the principal objective of regulating the workplace to prevent exposure above the limits, the old system of ICRP Publication 2 and the new system of ICRP Publications 26 and 30 are substantially equivalent.

The new ICRP system, however, lends itself to other applications not contemplated by the old system. The old system provided no explicit annual limits on intake of a radionuclide, or on accumulated dose from a deposited radionuclide—it provided only a limit on the attained dose rate (or radionuclide burden) after 50 years of continuous exposure. For radionuclides of short effective half-life, the 50th-year limit was effectively an annual limit, but for radionuclides of long effective half-life, it was not. If a worker under the old system incurred a plutonium deposition equal to one-half the permissible body burden, that worker was technically not exposed beyond the limits—neither the permissible body burden nor any annual organ dose equivalent (H_{50}) limit was exceeded. With the new system, the committed dose is to be attributed to the year of intake, and annual limits on intake (ALI) can be calculated based on these committed dose equivalents, as has been done in ICRP Publication 30. The ALI's are considered by the ICRP as secondary standards. They permit two kinds of action not possible under the old system—exposure may be categorized as excessive on an annual rather than on an attained basis; and, utilizing the other new concept of effective committed dose equivalent, dose from many external and internal sources may be added to give an administrative measure of total radiation exposure. It is in this area of retrospective evaluation of exposures that the two systems differ most and where controversy exists as to their relative merits.

The old system was as precise as it needed to be, considering the lack of precision in most of the biological parameters available for its implementation. It seems unlikely that a different system, even though more precise and conceptually more acceptable, will result in significantly better protection of workers. Both systems are inherently conservative.

The NCRP, therefore, favors the adoption of a new system of radiation protection such as that embodied in the use of the committed effective dose equivalent concept, but with stringent reservations as to the appropriate applications of that system. Its use should be limited to prospective protection planning, as reflected in derived standards for control of the workplace environment, and to such administrative operations as may be required by regulatory bodies to evaluate compliance by the employer with regulatory standards. A committed effective dose equivalent system should specifically *not* be used as a measure of an individual worker's exposure status. Recognition of this restriction is essential if the system is not to be abused.

Exposure to internally deposited radionuclides is, in many respects, a different insult from that occasioned by irradiation from external sources. The dose from internal emitters is usually heterogeneously distributed in both space and time. It is often concentrated in a single organ or portion of an organ; the radiation dose may be extended at a variably dimishing rate over years or a lifetime. External exposure is apt to be more uniform in its distribution throughout the body, and of limited duration in time compared to most internal emitters of interest. Summing these two kinds of exposure may simplify administrative record-keeping; but as applied to evaluation of the adequacy of workplace controls, or of consequences to the exposed individual, such summing can only result in a loss of information.

Exceeding an ALI (as defined in terms of committed effective dose equivalent) is properly considered a failure of the employer's protection practices; but should *not* be considered an overexposure of the individual unless that individual's current annual radiation dose exceeds the limit set by the primary standard. The large variability of individual organ and body burdens even under nominally identical exposure conditions and the consequent uncertainty of an individual's exposure status without actual measurement puts action involving an employee's status in a category different from evaluation of protection practices.

A statement of "permissible body burden" or "permissible organ burden" (perhaps better expressed as "derived" body burden or "derived" organ burden) remains the most practical guide to the actual internal exposure status of the individual. It is the total quantity of radionuclide deposited that determines dose equivalent to the tissue (which is the focus of the primary standard); it is this quantity that can be and is most often measured or estimated in practical health physics evaluations, frequently by regulatory requirements for routine bioassay. It is this quantity that most directly correlates with possible health consequences to the individual. The establishment of organ radionuclide burdens equivalent to the dose limit set by the primary standard is a necessary component of a system for control of internal emitters, if that system is to serve as a useful guide in the evaluation of individual exposures.

It must be emphasized that many aspects of the committed dose equivalent system of ICRP Publication 30, as well as of the earlier systems of ICRP and NCRP, involve conservative assumptions designed to simplify application and to insure that errors will be on the side of safety. In retrospective evaluation of an individual's exposure status, such assumptions and approximations are not appropriate if better information is available; and in the case of any serious overexposure, better information pertinent to that specific case should always be sought. Thus, specific data relative to the chemical and physical form of the material ingested or inhaled may alter predictions as to its biological behavior; deposition, distribution, and retention parameters measured on the exposed person are preferable to general models; the age and sex of the exposed person are relevant factors, particularly in relation to possible genetic consequences. While the system established for control of radiation exposure will unavoidably influence the retrospective evaluation of individual exposures, such influence must never equate with routine application. Final evaluation must be a matter of informed medical judgement dissociated from any system of exposure control.

The NCRP accepts an approach to radionuclide exposure control based upon an evaluation of risk, and the equating of risk to dose, where necessary, as embodied in ICRP Publications 26 and 30. Important reservations exist regarding any application of the system to evaluation of individual exposures in retrospect, and it is proposed that a statement of "derived body (or organ) burden" be included in the system. Reservations with regard to more detailed aspects of the system described in ICRP Publications 26 and 30 have been discussed in the earlier sections of this report.

In addition to the reservations expressed in this report, the NCRP recognizes that regulatory constraints control the ultimate adoption of any new system of radiation protection in the USA. As a result, the MPC's in the current regulations and NCRP Report No. 22 (NCRP, 1959) are still valid until official action is taken by the appropriate authorities.

Appendix A

A Comparison of Single and Continuous Intake

The ICRP now bases exposure limits on the assumption of single rather than continuous intake during the work year. This change simplifies the mathematical expressions obtained as solutions to the differential equations for the metabolic models. Following the year in which the exposure occurred, the dose equivalent is practically the same for either intake pattern, as illustrated by the simple example treated below. Consequently, the ALIs, computed by the ICRP from H_{50}, the dose equivalent at 50 years (also called the committed dose equivalent), are essentially independent of the intake pattern during the year of exposure.

In this example, attention is focused on a single target organ in which absorption is assumed to be instantaneous or at a constant rate and loss due to biological processes and physical decay is assumed to follow a single exponential with effective half life, $T_{1/2}$, expressed in years. Fig. A.1 provides a schematic representation of this model.

Equations: For a single intake, the equation governing the activity in the organ, $A(t)$[5], is:

$$\frac{\mathrm{d}A_s(t)}{\mathrm{d}t} = -\lambda A_s(t)\,, \qquad \text{(A-1)}$$

where $\lambda = \ln 2/T_{1/2}$ and t = time in years. For continuous intake at a constant rate over a period of T years,

$$\frac{\mathrm{d}A_c(t)}{\mathrm{d}t} = \frac{A_0}{T} - \lambda A_c(t) \qquad 0 \le t \le T \quad \text{(A-2)}$$

$$\frac{\mathrm{d}A_c(t)}{\mathrm{d}t} = -\lambda A_c(t) \qquad t \ge T, \quad \text{(A-3)}$$

where A_0 is the total activity absorbed by the organ.

[5] The symbols $A_s(t)$ and $A_c(t)$ will be used when necessary to distinguish between single and continuous intake.

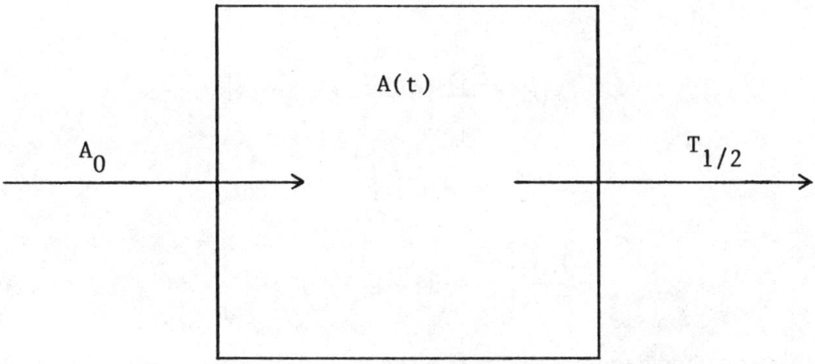

Fig. A.1 Metabolic model for single organ in which A_0 is the total uptake, $A(t)$ is the activity in the organ, and $T_{1/2}$ is the effective half period for loss by biological and physical processes.

The solutions to these equations are:

$$A_s(t) = A_0 e^{-\lambda t} , \qquad \text{(A-4)}$$

$$A_c(t) = \frac{A_0}{\lambda T} (1 - e^{-\lambda t}) , \qquad 0 \le t \le T \quad \text{(A-5)}$$

$$A_c(t) = \frac{A_0}{\lambda T} (1 - e^{-\lambda T}) e^{-\lambda (t-T)} . \qquad t \ge T \quad \text{(A-6)}$$

By virtue of their definitions, the dose equivalent rate to the organ, $\dot{H}(t)$, and activity in the organ are directly proportional to one another, i.e., $\dot{H}(t) = KA(t)$, where K is the constant of proportionality. Therefore, the dose equivalent, $H(t)$, can be determined by multiplying Eqs. (A-4) to (A-6) by K and integrating with respect to time. The dose equivalents are:

$$H_s(t) = \frac{KA_0}{\lambda} (1 - e^{-\lambda t}) , \qquad \text{(A-7)}$$

$$H_c(t) = \frac{KA_0}{\lambda T} \left[t - \frac{1}{\lambda} (1 - e^{-\lambda t}) \right] . \qquad 0 \le t \le T \quad \text{(A-8)}$$

For $t \ge T$, $H_c(t)$ is the sum of the dose equivalent at T, from Eq. (A-8), plus the dose equivalent following T, from the integration of Eq.

(A-6), that is,

$$H_c(t) = \frac{KA_0}{\lambda T}\left[T - \frac{1}{\lambda}(1 - e^{-\lambda T})\right]$$

$$+ \frac{KA_0}{\lambda T}(1 - e^{-\lambda T}) \int_T^t e^{-\lambda(t'-T)}\, dt'$$

$$= \frac{KA_0}{\lambda T}\left[T - \frac{1}{\lambda}(1 - e^{-\lambda T})\, e^{-\lambda(t-T)}\right]. \qquad t \geq T \quad \text{(A-9)}$$

In the above equations, $t = 0$ is the instant of single intake and also the start of the period of continuous intake. A comparison between the two intake patterns is more meaningful if $t = 0$ occurs at the midpoint of the continuous intake period rather than at its start. The shift in time origin can be achieved by replacing t in the equations for continuous intake by $t + T/2$. Then Eqs. (A-5) and (A-6) become:

$$A_c(t) = \frac{A_0}{\lambda T}[1 - e^{-\lambda(t+T/2)}], \quad -T/2 \leq t \leq T/2 \quad \text{(A-10)}$$

$$A_c(t) = \frac{A_0}{\lambda T}(1 - e^{-\lambda T})e^{-\lambda(t-T/2)}, \quad t \geq T/2 \quad \text{(A-11)}$$

and Eqs. (A-8) and (A-9) become:

$$H_c(t) = \frac{KA_0}{\lambda T}\left\{t + T/2 - \frac{1}{\lambda}[1 - e^{-\lambda(t+T/2)}]\right\}, \quad -T/2 \leq t \leq T/2 \quad \text{(A-12)}$$

$$H_c(t) = \frac{KA_0}{\lambda T}\left[T - \frac{1}{\lambda}(1 - e^{-\lambda T})e^{-\lambda(t-T/2)}\right]. \quad t \geq T/2 \quad \text{(A-13)}$$

In the ICRP system, H_{50} represents the dose equivalent at 50 years following a single intake, that is (Eq. (A-7) at 50 years):

$$H_{50} = \frac{KA_0}{\lambda}(1 - e^{-50\lambda}). \quad \text{(A-14)}$$

When Eq. (A-14) is solved for K and the expression is substituted into the equations for dose equivalent, Eqs. (A-7), (A-12), and (A-13) respectively, one obtains:

$$H_s(t) = \frac{H_{50}(1 - e^{-\lambda t})}{1 - e^{-50\lambda}}, \tag{A-15}$$

$$H_c(t) = \frac{H_{50}}{T(1 - e^{-50\lambda})}\left\{t + T/2 - \frac{1}{\lambda}[1 - e^{-\lambda(t+T/2)}]\right\}, \quad -T/2 \le t \le T/2 \tag{A-16}$$

$$H_c(t) = \frac{H_{50}}{T(1 - e^{-50\lambda})}\left[T - \frac{1}{\lambda}(1 - e^{-\lambda T})e^{-\lambda(t-T/2)}\right]. \quad t \ge T/2 \tag{A-17}$$

Figs. A.2a and A.2b compare the normalized activity, $A(t)/A_0$, and the normalized dose equivalent, $H(t)/H_{50}$, following constant continuous intake for one year and single intake at mid-year. The effective half life for loss, $T_{1/2} = 1$ year. The difference between the two intake patterns is hardly distinguishable after the end of the year of exposure. The results for other values of $T_{1/2}$ are quite similar as can be verified by plotting the relevant expressions.

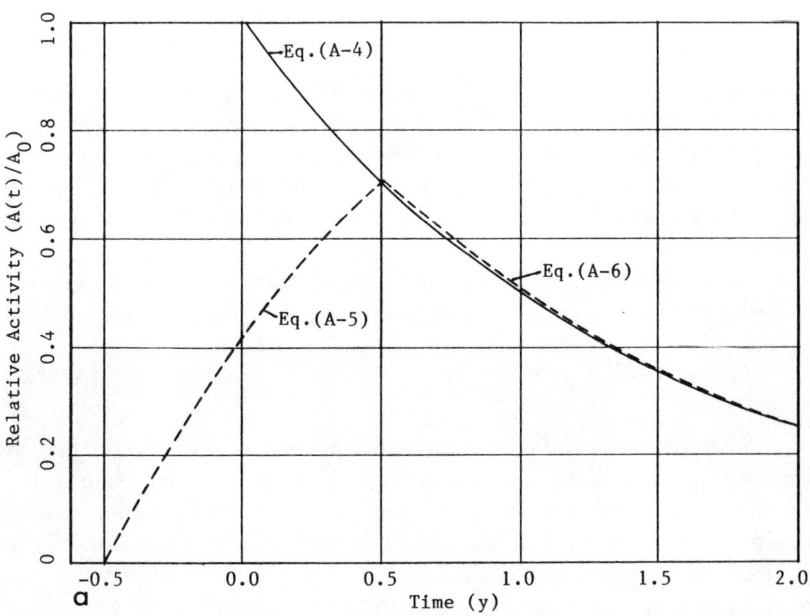

Fig. A.2a Comparison of organ activity for constant continuous intake during a 1-year period (– – – –) and for single intake at midyear (———).

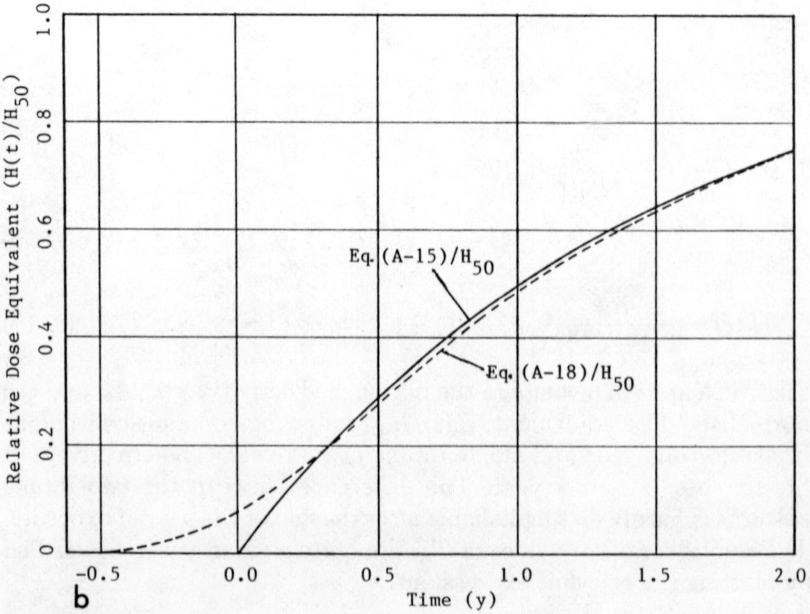

Fig. A.2b Comparison of dose equivalent accumulated by the organ during a 1-year period (– – – –) and for single intake at midyear (———).

Timing of the instant of Intake: If single intake were to occur at a time other than mid-year, the curves for single and continuous intake would lie farther apart. However, 50 years after single intake, the dose equivalent values for the two intake patterns deviate insignificantly regardless of when single intake occurs during the year of continuous exposure.

This conclusion stems from a modification of Eq. (A-9) in which t is replaced by $t + \tau$ and K is replaced by its solution from Eq. (A-14):

$$H_c(t) = \frac{H_{50}}{T(1-e^{-50\lambda})}\left[T - \frac{1}{\lambda}(1-e^{-\lambda T})e^{-\lambda(t-T+\tau)}, \quad t \geq T-\tau \right. \qquad \text{(A-18)}$$

where τ represents the amount of time after the start of continuous intake that single intake occurs. τ may take on any value from 0 to 1 year. $\tau = 0$ implies that single intake occurs when continuous intake begins, and $\tau = 1$ implies that single intake occurs at the end of the year of continuous intake. The fractional difference, f_0, between the dose equivalents for single and continuous intake at $t = 50$ years is:

$$f_0 = \frac{H_s(50) - H_c(50)}{H_{50}}. \qquad \text{(A-19)}$$

TABLE A.1—*Fractional difference, f_0, between H_s (50) and H_c (50)*

$T_{1/2}$ (y)	$\tau = 0$	$\tau = 0.5$ yr	$\tau = 1$ yr
1	$4. \times 10^{-16}$	$2. \times 10^{-17}$	$-2. \times 10^{-16}$
3	$1. \times 10^{-6}$	$2. \times 10^{-8}$	$-1. \times 10^{-16}$
10	$1. \times 10^{-3}$	$6. \times 10^{-6}$	$-1. \times 10^{-3}$
30	$5. \times 10^{-3}$	$1. \times 10^{-5}$	$-5. \times 10^{-3}$
100	$8. \times 10^{-3}$	$5. \times 10^{-6}$	$-8. \times 10^{-3}$
300	$9. \times 10^{-3}$	$2. \times 10^{-6}$	$-9. \times 10^{-3}$
1000	$1. \times 10^{-2}$	$2. \times 10^{-7}$	$-1. \times 10^{-2}$
10000	$1. \times 10^{-2}$	$-7. \times 10^{-5}$	$-1. \times 10^{-2}$

Because $H_s(50) = H_{50}$ [from Eq. (A-15)], then

$$f_0 = 1 - \frac{H_c\ (50)}{H_{50}}. \qquad \text{(A-20)}$$

Values of f_0 are given in Table A.1 for $T = 1$ year and for various values of $T_{1/2}$ and τ. Depending on the conditions, $H_s(50)$ may be greater or less than $H_c(50)$. However, the largest entry in the table, 0.01, is far too small to have any practical significance for radiation protection.

The foregoing example is based, in part, on instantaneous absorption by an organ consisting of one compartment. The ICRP model for most organs is a single compartment but instantaneous absorption cannot occur because of the finite time delays which occur as the radionuclide passes through the transfer compartment which precedes most organs. Clearance from the transfer compartment occurs with a half life of 0.25 day, which is short compared with the residence times for most nuclides in most organs. Thus, despite transfer compartment delays, absorption under most circumstances is approximated quite accurately by the instantaneous absorption model.

Appendix B-1

Excerpt Relating to Specific Effective Energy from ICRP Publication 30, *Limits for Intakes of Radionuclides by Workers*[6]

4.5. Specific Effective Energy (SEE)

In the dosimetric data for individual radionuclides given in the Supplement to this report, values are given for

$$\text{SEE}(T \leftarrow S) = \textstyle\sum_i \text{SEE}(T \leftarrow S)_i$$

for a number of target and source organs. It is emphasized that the values shown refer only to the radionuclide concerned and do not include any contribution from daughter radionuclides. Values for daughter radionuclides are given separately. For any radionuclide, j, $\text{SEE}(T \leftarrow S)_j$ for target T and source S is given by

$$\text{SEE}(T \leftarrow S)_j = \sum_i \frac{Y_i E_i AF(T \leftarrow S)_i Q_i}{M_T} \text{ MeV g}^{-1} \text{ per transformation,} \quad (4.8)$$

where the summation is over all radiations produced per transformation of radionuclide j in source organ S;

Y_i is the yield of radiations of type i per transformation of radionuclide j;

E_i(in MeV) is the average or unique energy of radiation i as appropriate;

[6] Excerpts are reproduced, with permission of the ICRP, section, table, figure, equation numbers, and references are as in the original. References from ICRP Publication 30 are not repeated here; for these, the original text should be consulted (ICRP, 1979a). For later statement on neutrons see (ICRP, 1985).

$AF(T \leftarrow S)_i$ is the fraction of energy absorbed in target organ T per emission of radiation i in S. For most organs it is assumed that the energies from alpha particles and electrons are completely absorbed within the source organ. Notable exceptions are mineral bone (see Chapter 7) and the contents of the gastrointestinal tract (see Chapter 6). The absorbed fraction of energy from photons is estimated by the use of data on specific absorbed fraction (absorbed fraction per g of target) given in *ICRP Publication 23.* The absorbed fractions for fission neutrons have been obtained from data given by Dillman and Jones (1975), and Ford *et al.* (1977);

Q_i is the quality factor appropriate for radiation of type i as discussed above in Section 4.3; and

M_T(in g) is the mass of the target organ.

Appendix B-2

Excerpt Relating to Committed Dose Equivalent from ICRP Publication 30, *Limits for Intakes of Radionuclides by Workers*[7]

4.3. Committed Dose Equivalent (H_{50})

For purposes of planning in radiological protection it is assumed that risk of a given biological effect is linearly related to dose equivalent. In these circumstances, risk of an effect is determined by the total dose equivalent averaged throughout the organ or tissue at risk, independent of the time over which that dose equivalent is delivered. For planning work with radioactive materials the Commission recommends that the appropriate period for integration of dose equivalent is a working life-time of 50 years. The total dose equivalent averaged throughout any tissue over the 50 years after intake of a radionuclide into the body is termed the committed dose equivalent, H_{50}, which is therefore given by

$$H_{50} = \sum_i \frac{\int^M D_{50,i} Q_i N_i \, \mathrm{d}m}{\int^M \mathrm{d}m} \qquad (4.1)$$

where M is the mass of the specified organ or tissue; and, for each type of radiation i;

$D_{50,i}$ is the total absorbed dose during a period of 50 years after

[7] Excerpts are reproduced, with permission of the ICRP; section, table, figure, equation numbers, and references are as in the original. References from ICRP Publication 30 are not repeated here; for these, the original text should be consulted (ICRP, 1979a).

intake of the radionuclide into the body in the element of mass dm of the specified organ or tissue;

Q_i is the quality factor, and

N_i is the product of all other modifying factors such as dose rate, fractionation, etc.

The quality factor, Q, is defined as a continuous function of collision stopping power in water (*ICRP Publication 21*). Therefore the value of Q_i will vary along the track of an ionizing particle and may be different for each element of mass dm in the irradiated tissue concerned. However, in view of the many uncertainties in estimating the dose to a tissue following the intake of a radioactive material, the Commission recommends (para. 20, *ICRP Publication 26*) that for internal exposure the value of Q for a given type of radiation may be considered constant and have one of three values as follows

$Q = 1$ for beta particles, electrons and all electromagnetic radiation including gamma radiation, x rays and bremsstrahlung.

$Q = 10$ for fission neutrons emitted in spontaneous fission and for protons.

$Q = 20$ for alpha particles from nuclear transformations, for heavy recoil particles and for fission fragments.

The Commission recommends that the product of all other modifying factors, N, should be taken as 1 for values of dose equivalent less than or equal to the limits specified in Chapter 2. (Para. 18, *ICRP Publication 26.*) It will not be considered further in this report.

Therefore, using the values of Q shown above, which are constant for any type of radiation i, the expression for H_{50} shown in eqn (4.1) simplifies to

$$H_{50} = \sum_i Q_i \overline{D_{50,i}} \qquad (4.2)$$

where $\overline{D_{50,i}}$ is the total absorbed dose during the 50 years after the intake of the radionuclide into the body averaged throughout the specified organ or tissue for each radiation of type i. A specified organ or tissue is one of those designated in Chapter 2 and in the metabolic data for individual elements.

In this report estimates are made of the committed dose equivalents in a number of target organs from the activity in a given source organ. For each type of radiation i, $H_{50,i}$ in target organ T resulting from radionuclide j in source organ S is the product of two factors

(a) the total number of transformations of radionuclide j in S over a period of 50 years after intake,

(b) the energy absorbed per g in T, suitably modified for quality factor, from radiation of type i per transformation of radionuclide j in S.

i.e. for each radiation of type i from radionuclide j

$$H_{50}(T \leftarrow S)_i = Q_i\overline{D_{50}}(T \leftarrow S)_i$$

$$= U_S \times 1.6 \times 10^{-13}\mathrm{SEE}(T \leftarrow S)_i \times 10^3 \ \mathrm{Sv}$$

where U_s is the number of transformations of j in S over the 50 years following intake of the radionuclide;

1.6×10^{-13} is the number of joules in 1 MeV;

SEE $(T \leftarrow S)_i$ (in MeV g^{-1} per transformation) is the specific effective energy for radiation type i, suitably modified by quality factor, absorbed in T from each transformation in S (see Section 4.5) and 10^3 is the conversion factor from g^{-1} to kg^{-1}.

$$\therefore H_{50}(T \leftarrow S)_i = 1.6 \times 10^{-10} U_S \mathrm{SEE}(T \leftarrow S)_i \quad \mathrm{Sv} \quad (4.3)$$

and for all types of radiation emitted by radionuclide j:

$$H_{50}(T \leftarrow S)_j = 1.6 \times 10^{-10}[U_S \textstyle\sum_i \mathrm{SEE}(T \leftarrow S)_i] \quad \mathrm{Sv}. \quad (4.4)$$

When the radionuclide has a radioactive daughter j'

$$H_{50}(T \leftarrow S)_{j+j'} = 1.6 \times 10^{-10}[\{U_S \textstyle\sum_i \mathrm{SEE}(T \leftarrow S)_i\}_j + \{U_S \textstyle\sum_i \mathrm{SEE}(T \leftarrow S)_i\}_{j'}] \quad \mathrm{Sv}. \quad (4.5)$$

In general, for the intake of any mixture of radionuclides, i.e. parent with daughters and/or other radionuclides, H_{50} in target T from activity in source S is given by

$$\textstyle\sum_j H_{50}(T \leftarrow S)_j = 1.6 \times 10^{-10} \textstyle\sum_j [U_S \textstyle\sum_i \mathrm{SEE}(T \leftarrow S)_i]_j \quad \mathrm{Sv} \quad (4.6)$$

where the summation in j is over all the radionuclides involved. Finally, target T may be irradiated by radiations arising in several different sources S. The total value of H_{50} in target T is then given by

$$H_{50,T} = 1.6 \times 10^{-10} \textstyle\sum_S \textstyle\sum_j [U_S \textstyle\sum_i \mathrm{SEE}(T \leftarrow S)_i]_j \quad \mathrm{Sv}. \quad (4.7)$$

Appendix B-3

Excerpt Relating to Dose Equivalent Limits, Weighting Factors and Stochastic and Non-Stochastic Effects for Occupational Exposure from ICRP Publication 30, *Limits for Intake of Radionuclides by Workers*[8]

2.1. Introduction

In the decade or so since the Commission's former recommendations were published, more information has become available on the relation between absorbed doses of ionizing radiation and risks of biological effects. This new information enables the Commission to establish limits for exposure to ionizing radiation on a different basis than it has used heretofore (*ICRP Publication* 26). Two broad categories of radiation-induced effects are considered, namely,

(i) malignant and hereditary disease for which the probability of an effect occurring, rather than its severity, is regarded as a function of dose without threshold (stochastic effects), and

[8] Excerpts are reproduced, with permission of the ICRP; section, table, figure, equation numbers, and references are as in the original. References from ICRP Publication 30 are not repeated here; for these, the original text should be consulted (ICRP, 1979a).

(ii) effects such as opacity of the lens and cosmetically unacceptable changes in the skin for which a threshold or pseudo-threshold of dose must be exceeded before the effect is induced (non-stochastic effects).

The Commission's recommendations are intended to prevent non-stochastic effects and to limit the occurrence of stochastic effects to an acceptable level.

2.2. Dose-Equivalent Limits for Occupational Exposure

The Commission believes that non-stochastic effects will be prevented by applying a dose-equivalent limit of 0.5 Sv (50 rem) in a year to all tissues except the lenses of the eyes, for which the Commission recommends a limit of 0.3 Sv (30 rem) in a year. These limits apply irrespective of whether the tissues are exposed singly or together with other organs, and they are intended to constrain exposures that fulfil the limitation on stochastic effects discussed below.

For stochastic effects the Commission's recommended system for limiting exposure is based on the principle that the limit on risk should be equal whether the whole body is irradiated uniformly or whether there is non-uniform irradiation. This condition will be met if

$$\sum w_T H_T \leq H_{wb} \tag{2.1}$$

where w_T is a weighting factor representing the ratio of the stochastic risk resulting from tissue (T) to the total risk when the whole body is irradiated uniformly;

H_T is the dose equivalent received by tissue (T), and

H_{wb} is the stochastic dose-equivalent limit for uniform irradiation of the whole body.

For occupational exposure the Commission recommends the value 50 mSv (5 rem) in any year for H_{wb} and the values of w_T shown in Table 2.1. The values are recommended as appropriate for the protection of any worker, independent of age or sex (para. 104–106, *ICRP Publication 26*).

The value of w_T shown for bone surfaces relates to endosteal cells and to the epithelium on bone surfaces (Chapter 7). The value of w_T for the remainder of tissues requires further clarification. For reasons stated by the Commission in para. 105, *ICRP Publication 26*, a value of w_T of 0.06 is applicable to each of the five organs or tissues of this

Table 2.1. Weighing factors recommended by the Commission for stochastic risks

Organ or tissue	w_T
Gonads	0.25
Breast	0.15
Red bone marrow	0.12
Lung	0.12
Thyroid	0.03
Bone surfaces	0.03
Remainder	0.30

remainder receiving the greatest dose equivalents and the exposure of all other tissues in this group is neglected.

Because of their very low sensitivity to cancer induction, skin and lens (paras. 61–64, *ICRP Publication 26)* are not considered as part of remainder tissue for limiting stochastic effects. When the gastrointestinal tract is irradiated, the stomach, small intestine, upper large intestine and lower large intestine are considered as four separate organs.

2.3. Limits for the Intake of Radioactive Materials by Workers

The estimates of risk of radiation-induced cancer and hereditary disease on which the Commission's dose-equivalent limits for stochastic effects are based were made using the hypothesis that risk of an effect is linearly related to dose equivalent. Therefore, it is the total dose equivalent averaged throughout any organ or tissue, independently of the time over which that dose equivalent is delivered, which determines the degree of effect in that tissue. With regard to limits on the intake of a radioactive material into the body, the Commission has reconsidered the question of the time over which this total dose equivalent should be integrated and has concluded that the period of 50 years used heretofore is appropriate for an occupational lifetime. The total dose equivalent in any tissue over the 50 years after intake of a radionuclide into the body is termed the Committed Dose Equivalent, H_{50}. It is emphasized that this is the dose equivalent which a Reference Man is assumed to receive if he lives for 50 years after his intake of the radioactive material and if no steps are taken to accelerate the removal of the radionuclide from his body.

Therefore, in order to meet the Commission's basic limits for the exposure of workers, the intakes of radioactive materials in any year must be limited to satisfy the following conditions

$$\sum_T w_T H_{50,T} \leq 0.05 \qquad \text{Sv} \qquad (2.2a)$$

and

$$H_{50,T} \leq 0.5 \qquad \text{Sv} \qquad (2.2b)$$

where w_T is the weighting factor in Table 2.1, and

$H_{50,T}$ (in Sv) is the total committed dose equivalent in tissue (T) resulting from intakes of radioactive materials from all sources during the year in question.

Relationship (2.2a) limits stochastic effects and (2.2b) non-stochastic effects, arising from intakes of radioactive materials. With regard to (2.2b), the limit for non-stochastic effects in any tissue is taken as 0.5 Sv (50 rem), since no case is known where lens opacity would be the factor limiting intake of radioactive materials. It might be the limiting factor when the body is irradiated from the exterior by submersion in a radioactive noble gas and values for the derived air concentration in such cases are discussed in Chapters 3 and 8.

Appendix B-4

Excerpt Relating to Lung Model from ICRP Publication 30 *Limits for Intakes of Radionuclides by Workers*[9]

The respiratory system is divided into three distinct regions—the nasal passage (N-P), the trachea and bronchial tree (T-B) and the pulmonary parenchyma (P). Deposition is assumed to vary with the aerodynamic properties of the aerosol distribution and is described by the three parameters $D_{N\text{-}P}$, $D_{T\text{-}B}$, and D_P which represent the fractions of inhaled material initially deposited in the N-P, T-B, and P regions, respectively. For a log-normal distribution of diameters, which seems typical of aerosols, the pattern of deposition can be related to the activity median aerodynamic diameter (AMAD) of the aerosol (Fig. 5.1). In this report calculations of committed dose equivalent are for an aerosol with an AMAD of 1 μm. Estimates for other AMADs can be made using the data in Fig. 5.1 as discussed in Section 5.5 below.

To describe the clearance of inhaled radioactive materials from the lung, materials are classified as D, W or Y which refer to their retention in the pulmonary region. This classification applies to a range of half-times for D of less than 10 days, for W from 10 to 100 days and for Y greater than 100 days. The three regions, N-P, T-B and P described above are each divided into two or four compartments as shown in Fig. 5.2. Each of these compartments is associated with a particular pathway of clearance for which the half-time of clearance is T days and the fraction leaving the region by that rate is F. Thus, compart-

[9] Excerpts are reproduced, with permission of the ICRP; section, table, figure, equation numbers, and references are as in the original. References from ICRP Publication 30 are not repeated here; for these, the original text should be consulted (ICRP, 1979a). Figures 5.1 and 5.2 cited herein are shown in Section 5 of this report.

ments a, c and e are associated with absorption processes, whereas b, d, f and g are associated with particle transport processes, including mucociliary transport, which translocate material to the gastrointestinal tract. The pulmonary lymphatic system (L) also serves to remove dust from the lungs. It is associated with compartment h in the P region of the lungs from which material is translocated to compartments i and j in the pulmonary lymph nodes. Material in compartment i is translocated to body fluids but that in compartment j is assumed to be retained there indefinitely. This compartment is only considered appropriate for class Y aerosols; for class D and W aerosols the fraction of material entering compartment j from compartment h (F_j) is set equal to zero. The clearance of material from each of the compartments described above is assumed to be governed by first order kinetics so that each compartment is associated with a clearance constant λ and half-time of clearance $T = 0.693/86400\lambda$. The clearance of inhaled material from the lung is, therefore, described by a set of interlinked first order differential equations, as follows

$$\frac{\mathrm{d}}{\mathrm{d}t} q_{\mathrm{a}}(t) = I(t).\ D_{\mathrm{N\text{-}P}}.\ F_{\mathrm{a}} - \lambda_{\mathrm{a}} q_{\mathrm{a}}(t) - \lambda_R q_{\mathrm{a}}(t) \qquad (5.1\mathrm{a})$$

$$\frac{\mathrm{d}}{\mathrm{d}t} q_{\mathrm{b}}(t) = I(t).\ D_{\mathrm{N\text{-}P}}.\ F_{\mathrm{b}} - \lambda_{\mathrm{b}} q_{\mathrm{b}}(t) - \lambda_R q_{\mathrm{b}}(t) \qquad (5.1\mathrm{b})$$

$$\frac{\mathrm{d}}{\mathrm{d}t} q_{\mathrm{c}}(t) = I(t).\ D_{\mathrm{T\text{-}B}}.\ F_{\mathrm{c}} - \lambda_{\mathrm{c}} q_{\mathrm{c}}(t) - \lambda_R q_{\mathrm{c}}(t) \qquad (5.1\mathrm{c})$$

$$\frac{\mathrm{d}}{\mathrm{d}t} q_{\mathrm{d}}(t) = I(t).\ D_{\mathrm{T\text{-}B}}.\ F_{\mathrm{d}} + \lambda_f q_f(t) + \lambda_{\mathrm{g}} q_{\mathrm{g}}(t) - \lambda_{\mathrm{d}} q_{\mathrm{d}}(t) - \lambda_R q_{\mathrm{d}}(t) \qquad (5.1\mathrm{d})$$

$$\frac{\mathrm{d}}{\mathrm{d}t} q_{\mathrm{e}}(t) = I(t).\ D_{\mathrm{P}}.\ F_{\mathrm{e}} - \lambda_{\mathrm{e}} q_{\mathrm{e}}(t) - \lambda_R q_{\mathrm{e}}(t) \qquad (5.1\mathrm{e})$$

$$\frac{\mathrm{d}}{\mathrm{d}t} q_{\mathrm{f}}(t) = I(t).\ D_{\mathrm{P}}.\ F_{\mathrm{f}} - \lambda_{\mathrm{f}} q_{\mathrm{f}}(t) - \lambda_R q_{\mathrm{f}}(t) \qquad (5.1\mathrm{f})$$

$$\frac{\mathrm{d}}{\mathrm{d}t} q_{\mathrm{g}}(t) = I(t).\ D_{\mathrm{P}}.\ F_{\mathrm{g}} - \lambda_{\mathrm{g}} q_{\mathrm{g}}(t) - \lambda_R q_{\mathrm{g}}(t) \qquad (5.1\mathrm{g})$$

$$\frac{\mathrm{d}}{\mathrm{d}t} q_{\mathrm{h}}(t) = I(t).D_{\mathrm{P}}.\ F_{\mathrm{h}} - \lambda_{\mathrm{h}} q_{\mathrm{h}}(t) - \lambda_R q_{\mathrm{h}}(t) \qquad (5.1\mathrm{h})$$

$$\frac{\mathrm{d}}{\mathrm{d}t} q_i(t) = F_i \lambda_{\mathrm{h}} q_{\mathrm{h}}(t) - \lambda_i q_i(t) - \lambda_R q_i(t) \qquad (5.1\mathrm{i})$$

$$\frac{d}{dt} q_j(t) = F_j \lambda_h q_h(t) - \lambda_R q_j(t) \tag{5.1j}$$

where $q_a(t)$, $q_b(t)$, etc. are the activities of an inhaled radionuclide in compartment a, b, etc. at time t;

$I(t)$ is the rate of inhalation of activity of the radionuclide;
λ_a to λ_i are the biological clearance rates of compartments a to i;
λ_R is the radioactive decay constant of the radionuclide, and
F_a to F_j are the fractions of material entering the various regions of the lung that are associated with the various compartments contained therein.

Values of F_a to F_j are given in Fig. 5.2 for the various inhalation classes D, W and Y. Also given are T_a to T_i for these various classes. λ_a to λ_i are obtained from the values of T_a to T_i by use of the formula $\lambda = 0.693/86400\,T$,

where λ is in s^{-1};

T is in days, and
86400 is the number of seconds in a day.

The classification D, W or Y for different chemical forms of a radionuclide are given in the metabolic data for the individual elements. As stated in Chapter 4 it is assumed that radioactive daughters remain with and behave metabolically like the inhaled parent radionuclide. In fact, there is a paucity of evidence about the behaviour of radioactive daughters produced within the lung. However, it is at least plausible to assume that daughters produced within a radioactive particle stay within the particle. The build-up and clearance of a radiocative daughter produced in the different compartments of the lung is, therefore, governed by a series of first order differential equations similar to those given in eqn (5.1), If $q'(t)$ is the activity of the radioactive daughter in any compartment at time t and $q(t)$ is the activity of its immediate precursor, then

$$\frac{d}{dt} q_a'(t) = \lambda_R' q_a(t) - \lambda_a q_a'(t) - \lambda_R' q_a'(t) \tag{5.2a}$$

$$\frac{d}{dt} q_b'(t) = \lambda_R' q_b(t) - \lambda_b q_b'(t) - \lambda_R' q_b'(t) \tag{5.2b}$$

$$\frac{d}{dt} q_c'(t) = \lambda_R' q_c(t) - \lambda_c q_c'(t) - \lambda_R' q_c'(t) \tag{5.2c}$$

$$\frac{d}{dt} q_d'(t) = \lambda_R' q_d(t) + \lambda_f q_f(t) \tag{5.2d}$$

$$+ \lambda_g q_g'(t) - \lambda_d q_d'(t) - \lambda_R' q_d'(t)$$

$$\frac{d}{dt} q_e'(t) = \lambda_R' q_e(t) - \lambda_e q_e'(t) - \lambda_R' q_e'(t) \tag{5.2e}$$

$$\frac{d}{dt} q_f'(t) = \lambda_R' q_f(t) - \lambda_f q_f'(t) - \lambda_R' q_f'(t) \tag{5.2f}$$

$$\frac{d}{dt} q_g'(t) = \lambda_R' q_g(t) - \lambda_g q_g'(t) - \lambda_R' q_g'(t) \tag{5.2g}$$

$$\frac{d}{dt} q_h'(t) = \lambda_R' q_h(t) - \lambda_h q_h'(t) - \lambda_R' q_h'(t) \tag{5.2h}$$

$$\frac{d}{dt} q_i'(t) = \lambda_R' q_i(t) + F_i \lambda_h q_h'(t) - \lambda_i q_i'(t) - \lambda_R' q_i'(t) \tag{5.2i}$$

$$\frac{d}{dt} q_j'(t) = \lambda_R' q_j(t) + F_j \lambda_h q_h'(t) - \lambda_R' q_j'(t) \tag{5.2j}$$

where λ_R' is the radioactive decay constant of a daughter radionuclide, and

F_i and F_j are the values appropriate to the class of inhaled parent radionuclide.

Similarly, a system of equations can be derived that describe the activities of a chain of parent and daughter radionuclides, the activity of each daughter being determined by its predecessor in the chain. The metabolic behaviour of all the radioactive progeny is assumed to be the same as that of the ancestral radionuclide which was inhaled.

5.3 Transfer of a Radionuclide from the Lungs Directly to Body Fluids or to the Gastrointestinal Tract

From Fig. 5.2 and the above equations (5.1a to 5.1j and 5.2a to 5.2j) it can easily be seen that the rate of transfer of a radionuclide directly from the lungs to body fluids (BF(t)), is given by

$$\mathrm{BF}(t) = \lambda_a q_a(t) + \lambda_c q_c(t) + \lambda_e q_e(t) + \lambda_i q_i(t) \tag{5.3}$$

whether the radionuclide is inhaled or produced in the lungs. Similarly the rate of transfer of activity of a radionuclide from the lungs to the gastrointestinal tract (G(t)), is given by

$$G(t) = \lambda_b q_b(t) + \lambda_d q_d(t) \tag{5.4}$$

whether the radionuclide is inhaled or produced in the lungs. The metabolism of a radionuclide after reaching either body fluids or the gastrointestinal tract following inhalation of its parent is usually assumed to be governed by the metabolic model for its inhaled parent (see Chapter 4).

The equations above completely specify the lung model used in this report and have been used to calculate the number of transformations (U) in the lung and other organs given in the dosimetric data for the various radionuclides which are collected in the Supplement to this report. Details of these calculations are not given here and the reader is referred to Appendix 1 for a brief discussion of methods that may be employed in estimating these quantities

5.4. Calculation of Committed Dose Equivalent, H_{50}, in the Lung

As described in Chapter 4, $H_{50,T}$ for a target organ T from a number of source organs S containing any mixture of radionuclides j is given by

$$H_{50,T} = 1.6 \times 10^{-10} \sum_S \sum_j [U_S \sum_i \mathrm{SEE}(T \leftarrow S)_i] \quad \mathrm{Sv} \tag{5.5}$$

where U_S is the number of transformations in the source organ S over 50 years following the intake of radionuclide j, and

$\mathrm{SEE}(T \leftarrow S)_i$(in MeV g^{-1} transformation^{-1}) is the specific effective energy absorbed in T from radiations of type i emitted in S.

The methods by which U_S can be calculated for the various parts of the respiratory system have been described in Section 5.2 both for the parent radionuclide and for its radioactive daughters. $\mathrm{SEE}(T \leftarrow S)_i$ for any radiation of type i emitted by radionuclide j is given by

$$\mathrm{SEE}(T \leftarrow S)_i = \frac{Y_i E_i AF(T \leftarrow S)_i Q_i}{M_T} \ \mathrm{MeV\ g^{-1}\ per\ transformation} \tag{5.6}$$

where Y_i is the yield of radiations of type i per transformation of radionuclide j;

E_i(in MeV) is the average, or unique, energy of radiation i;

$\mathrm{AF}(T \leftarrow S)_i$ is the average fraction of energy absorbed in target

organ T per emission of radiation i in source organ S;
Q_i is the quality factor appropriate for radiation i, as discussed in Chapter 4, and M_T (in g) is the mass of the target organ.

As explained in Section 5.1, dose in the N-P region is neglected and the target tissue T is eqn (5.5) is the lung comprising the T-B, P and L regions of total mass 1000 g. $H_{50,T}$ has two components—the committed dose equivalent from radioactive material in the lung, (i.e. in target tissue T comprising T-B, P and L) and the committed dose equivalent from photons arising in other organs and tissues of the body, S.

$$H_{50,T} = 1.6 \times 10^{-10} \, [\Sigma_j \, U_T{}^j \, \Sigma_i \, \mathrm{SEE}(T \leftarrow T)_i + \Sigma_S \, \Sigma_j \, U_S{}^j \, \Sigma_i \, \mathrm{SEE}(T \leftarrow S)_i] \quad \mathrm{Sv} \qquad (5.7)$$

In each case the summation is for all radionuclides, including those daughter radionuclides which build up after inhalation of their parent. Values of $U_T{}^j$ (= $U^j_{T\text{-B}} + U_P{}^j + U_L{}^j$) and corresponding values for daughters are given in the dosimetric data for parent radionuclide j. If they contribute significantly to dose, values for $U_S{}^j$ and corresponding values for daughters are also given in the dosimetric data for radionuclide j.

Values of SEE for any radionuclide j are derived from eqn (5.6) in which M_T is 1000 g and $\mathrm{AF}(T \leftarrow S)_i$ takes the following values:

(a) For $\mathrm{SEE}(T \leftarrow T)_i$, i.e. for radioactive material in the lung itself, $\mathrm{AF}(T \leftarrow T)_i = 1$ for all radiations other than photons and neutrons. For photons and fission neutrons values of $\mathrm{AF}(T \leftarrow T)_i/M_T$, the specific absorbed fraction, are either given in *ICRP Publication 23* or derived from data given by Dillman and Jones (1975) and Ford *et al.* (1977).

(b) For $\mathrm{SEE}(T \leftarrow S)_i$, i.e. for radioactive materials in organs and tissues S outside the respiratory system, $\mathrm{AF}(T \leftarrow S)_i = 0$ for all radiations other than photons and neutrons. For photons and fission neutrons values of $\mathrm{AF}(T \leftarrow S)_i/M_T$, the specific absorbed fraction, are given either in *ICRP Publication 23* or derived from data given by Dillman and Jones (1975) and Ford *et al.* (1977).

5.5 Particle Size Correction

Values of H_{50} per unit intake, annual limit on intake (ALI) and derived air concentration (DAC) given in the dosimetric data collected in the Supplement to this report for individual radionuclides are for a radionuclide with an AMAD of 1 μm. Together with values for H_{50}, values are given in parentheses for the fraction f_{N-P}, f_{T-B} and f_P of the committed dose equivalent in the reference tissue resulting from deposition in the N-P, T-B and P regions respectively. H_{50} for an aerosol of AMAD other than 1 μm may then be estimated by

$$\frac{H_{50}(\text{AMAD})}{H_{50}(1\,\mu\text{m})} = f_{N-P}\frac{D_{N-P}(\text{AMAD})}{D_{N-P}(1\,\mu\text{m})} + f_{T-B}\frac{D_{T-B}(\text{AMAD})}{D_{T-B}(1\,\mu\text{m})} + f_P\frac{D_P(\text{AMAD})}{D_P(1\,\mu\text{m})} \quad (5.8)$$

where D_{N-P}, D_{T-B} and D_P are the deposition probabilities in the respiratory regions for a given AMAD (Fig. 5.1).

If AMAD is unknown, it is recommended that values of H_{50} ALI and DAC given in the dosimetric data for an aerosol of 1 μm AMAD be used.

APPENDIX B-5

Excerpt Relating to Gastrointestinal Tract Model from ICRP Publication 30, *Limits for Intakes of Radionuclides by Workers*[10]

6.2. Dosimetric Model

The dosimetric model is based on the biological model developed by Eve (1966). For the purposes of radiological protection the GI tract is taken to consist of the 4 sections shown schematically in Fig. 6.1. (See p. 33).

Each of these sections is considered as a single compartment and translocation from one compartment to the next is taken to be governed by first order kinetics. Thus, if $q(t)$ is the activity of ingested radionuclide in a compartment at time t then the model is completely described by the following equations

$$\frac{\mathrm{d}}{\mathrm{d}t} q_{\mathrm{SI}}(t) = -\lambda_{\mathrm{SI}} q_{\mathrm{SI}}(t) - \lambda_R q_{\mathrm{SI}}(t) + \dot{\mathrm{I}}(t) \qquad (6.1a)$$

[10] Excerpts are reproduced, with permission of the ICRP section, table, figure, equation numbers, and references are as in the original. References from ICRP Publication 30 are not repeated here; for these, the original text should be consulted (ICRP, 1979a). Figure 6.1 cited herein is shown as Figure 5.3 of the text.

$$\frac{\mathrm{d}}{\mathrm{d}t} q_{\mathrm{SI}}(t) = -\lambda_{\mathrm{ST}} q_{\mathrm{SI}}(t) - \lambda_R q_{\mathrm{SI}}(t) - \lambda_{\mathrm{B}} q_{\mathrm{SI}}(t) + \lambda_{\mathrm{SI}} q_{\mathrm{SI}}(t) \tag{6.1b}$$

$$\frac{\mathrm{d}}{\mathrm{d}t} q_{\mathrm{ULI}}(t) = -\lambda_{\mathrm{ULI}} q_{\mathrm{ULI}}(t) - \lambda_R q_{\mathrm{ULI}}(t) + \lambda_{\mathrm{SI}} q_{\mathrm{SI}}(t) \tag{6.1c}$$

$$\frac{\mathrm{d}}{\mathrm{d}t} q_{\mathrm{LLI}}(t) = -\lambda_{\mathrm{LLI}} q_{\mathrm{LLI}}(t) - \lambda_{\mathrm{R}} q_{\mathrm{LLI}}(t) + \lambda_{\mathrm{ULI}} q_{\mathrm{ULI}}(t) \tag{6.1d}$$

where λ_R is the radioactive decay constant of the radionuclide in question;

$\lambda_{\mathrm{B}} q_{\mathrm{SI}}(t)$ is the rate of transfer of activity to body fluids from the small intestine, assumed to be the only site of absorption from the GI tract to body fluids, and
$I(t)$ is the rate of ingestion of activity of the radionuclide at time t.

Similarly, the model for a radioactive daughter produced in the GI tract from its ingested parent radionuclide is completely described by

$$\frac{\mathrm{d}}{\mathrm{d}t} q'_{\mathrm{SI}}(t) = -\lambda_{\mathrm{SI}} q'_{SI}(t) - \lambda_R' q'_{\mathrm{SI}}(t) + \lambda_R' q_{\mathrm{SI}}(t) \tag{6.2a}$$

$$\frac{\mathrm{d}}{\mathrm{d}t} q'_{\mathrm{SI}}(t) = -\lambda_{\mathrm{SI}} q'_{\mathrm{SI}}(t) - \lambda_R' q'_{\mathrm{SI}}(t) - \lambda_{\mathrm{B}} q'_{\mathrm{SI}}(t) + \lambda_{\mathrm{SI}} q'_{\mathrm{SI}}(t) + \lambda_R' q_{\mathrm{SI}}(t) \tag{6.2b}$$

$$\frac{\mathrm{d}}{\mathrm{d}t} q'_{\mathrm{ULI}}(t) = -\lambda_{\mathrm{ULI}} q'_{\mathrm{ULI}}(t) - \lambda_R' q'_{\mathrm{ULI}}(t) + \lambda_{\mathrm{SI}} q'_{\mathrm{SI}}(t) + \lambda_R' q_{\mathrm{ULI}}(t) \tag{6.2c}$$

$$\frac{\mathrm{d}}{\mathrm{d}t} q'_{\mathrm{LLI}}(t) = -\lambda_{\mathrm{LLI}} q'_{\mathrm{LLI}}(t) - \lambda_R' q'_{\mathrm{LLI}}(t) + \lambda_{\mathrm{ULI}} q'_{\mathrm{ULI}}(t) + \lambda_R' q_{\mathrm{LLI}}(t) \tag{6.2d}$$

where λ_R' is the radioactive decay constant of the radionuclide in question;

$q'(t)$ is its activity in any compartment and $q(t)$ is the activity of its immediate parent.

The value of λ_B can be estimated from f_1, the fraction of a stable element reaching the body fluids following ingestion:

$$\frac{\lambda_B}{\lambda_{SI} + \lambda_B} = f_1$$
$$\therefore \frac{f_1 \lambda_{SI}}{1 - f_1} = \lambda_B \tag{6.3}$$

Values of f_1 are given in the metabolic data for a number of classes of compounds of each individual element. For radioactive daughters, produced by decay of their parents in the GI tract, the value of f_1 used is usually that appropriate to the stable element of which the ingested radionuclide is an isotope (see Chapter 4). Where a value $f_1 = 1$ is given it is assumed that the radionuclide passes directly from the stomach to body fluids and does not pass through other sections of the gastrointestinal tract.

A system of equations similar to (6.1) and (6.2) can be derived that describe the activities of a chain of parent and daughter radionuclides, the activity of each daughter being determined by the activity of its predecessor in the chain. The metabolic behaviour of all the radioactive progeny is assumed to be the same as that of the ancestral radionuclide which was ingested.

6.3. Activity Transferred from the Respiratory System

The activity of an inhaled radionuclide in any section of the GI tract is described by eqns (6.1a)–(6.1d) in which the term $I(t)$ in eqn (6.1a) is the rate of entry of activity of the radionuclide to the GI tract from the respiratory system (Chapter 5). The activity of a daughter radionuclide in any section of the GI tract is described by eqns (6.2a)–(6.2d) with the addition of a term, $+ I'(t)$, to the right hand side of eqn (6.2a). $I'(t)$ is the rate of entry of activity of the daughter radionuclide to the GI tract, the daughter having been produced by decay of its parent in the respiratory system. The value of f_1 for any daughter produced in the respiratory system is usually assumed to be the same as that of the parent radionuclide (Chapter 4).

6.4. Activity Transferred to Body Fluids

The activity transferred to body fluids is the integral over 50 years of $\lambda_B q_{SI}(t)$ for the parent and $\lambda_B q'_{SI}(t)$ for its daughter. Their evaluation is described in the Appendix.

APPENDIX B-6.

Excerpt Relating to Bone Models from ICRP Publication 30, *Limits for Intakes of Radionuclides by Workers*[11]

7.1. Introduction

The cells at carcinogenic risk in the skeleton have been identified as the haematopoietic stem cells of marrow, and among the osteogenic cells, particularly those on endosteal surfaces, and certain epithelial cells close to bone surfaces (*ICRP Publication 11*). The haematopoietic stem cells in adults are assumed to be randomly distributed predominantly throughout the haematopoietic marrow within trabecular bone (*ICRP Publication 11*). Therefore, dose equivalent to those cells is calculated as the average over the tissue which entirely fills the cavities within trabecular bone. For the osteogenic tissue on endosteal surfaces and epithelium on bone surfaces the Commission recommends that dose equivalent should be calculated as an average over tissue up to a distance of 10 μm from the relevant bone surfaces (para. 47, *ICRP Publication 26*).

In order to estimate committed dose equivalent in these tissues from an intake of a particular radionuclide, it is necessary to know the distribution and retention of the radionuclide in the various tissues of the skeleton. At present such data are only available for a few radionuclides and it is hoped that further research in this field will be

[11] Excerpts are reproduced, with permission of the ICRP section, table, figure, equation numbers, and references are as in the original. References from ICRP Publication 30 are not repeated here; for these, the original text should be consulted (ICRP, 1979a).

encouraged. Meanwhile, for purposes of radiological protection, the methods described below are considered to be sufficiently accurate to estimate the committed dose equivalent to cells on bone surfaces and to active red bone marrow from intakes of all bone-seeking radionuclides.

7.2. Calculation of Committed Dose Equivalent, H_{50}, to Cells on Bone Surfaces and Active Red Bone Marrow

As explained in Chapter 4, H_{50} in a target organ T is given by

$$H_{50,T} = 1.6 \times 10^{-10} \textstyle\sum_S \sum_j [U_S \sum_i \mathrm{SEE}(T \leftarrow S)_i]_j \qquad \mathrm{Sv} \qquad (7.1)$$

where U_S is the number of transformations of nuclide j occurring in source organ S, and

$$\mathrm{SEE}(T \leftarrow S)_i = \frac{Y_i E_i \mathrm{AF}(T \leftarrow S)_i Q_i}{M_T} \ \mathrm{MeV\ g^{-1}\ transformation^{-1}} \qquad (7.2)$$

where Y_i is the yield of radiation of type i per transformation;

E_i (in MeV) is the average or unique energy of radiation i as appropriate;

$\mathrm{AF}(T \leftarrow S)_i$ is the fraction of energy absorbed in target organ T from radiation i originating in S;

Q_i is the quality factor appropriate for radiation of type i, and

M_T (in g) is the mass of the target organ.

In the case of bone the two target tissues are the cells near bone surfaces (BS) and the active red bone marrow (RM) as described above. For all radionuclides, except γ emitters, the source tissues will normally be cortical and trabecular bone.

Bone dosimetry, therefore, involves estimates of U_S for trabecular and cortical bone and estimates of $\mathrm{AF}(T \leftarrow S)_i$ for all radiations, radionuclides, source organs and target tissues.

From metabolic and dosimetric considerations radionuclides and their emissions are divided into the following six broad classes

(a) photons from all radionuclides;
(b) α particles from radionuclides assumed to be uniformly distributed throughout the volume of bone;
(c) α particles from radionuclides assumed to be on bone surfaces;
(d) β particles from radionuclides assumed to be uniformly distrib-

uted throughout the volume of bone;

(e) β particles from radionuclides assumed to be on bone surfaces and having an average β energy ($\bar{E}_\beta$) greater than 0.2 MeV, and
(f) β particles from radionuclides assumed to be on bone surfaces for which $\bar{E}_\beta$ is less than 0.2 MeV.

It should be noted that a radionuclide can belong to more than one class; for example, it may be a β emitter with γ-decay mode. In such a case the dose from each different type of radiation must be considered separately. This is why $AF(T \leftarrow S)_i$ is defined for each particular emission from each particular radionuclide. It should also be noted that a clear distinction is made between the dosimetry for radionuclides, such as ^{226}Ra, from which most transformations occur whilst they are distributed throughout the volume of bone, and others, such as ^{239}Pu, which tend to be bound to bone surfaces, although they may subsequently be buried by bone deposition (Vaughan, 1973; Marshall and Lloyd, 1973). The decision as to whether a radionuclide is assumed to be distributed throughout bone volume or assumed to be on bone surfaces is dealt with in the metabolic data for the various elements. However, two broad criteria can be listed here.

(1) Isotopes of the alkaline earth elements with radioactive half-lives greater than 15 days are considered to be uniformly distributed throughout the volume of bone (*ICRP Publication 20*).
(2) Radionuclides with radioactive half-lives of less than 15 days are considered to be distributed on bone surfaces, since they are unlikely to move far into the volume of bone before they decay.

Thus ^{224}Ra is assumed to be on bone surfaces and ^{226}Ra is assumed to be uniformly distributed throughout the volume of bone.

For the purposes of dosimetry a simplified specification of the dimensions and location of the two target tissues in bone has been adopted. Thus, for cells near bone surfaces (BS) the average committed dose equivalent is calculated for a layer of tissue 10 μm thick covering all endosteal surfaces and surfaces lined with epithelium. From the work of Lloyd and Hodges (1971) the total area of the endosteal surfaces in man has been estimated to be 11.2 m^2. However, more recent work by Beddoe *et al.* (1976) suggests a value of about 16.0 m^2 (Spiers and Vaughan, 1976). By comparison the area of epithelium on bone is small. In this report the total endosteal area of the skeleton has been taken to be 12 m^2, half being associated with cortical bone and half with trabecular bone (*ICRP Publications 20* and *23*). Therefore, the 10 μm thick layer on bone surfaces over which committed dose equivalent is to be averaged (BS) has a mass of 120 g. The mass of active red bone marrow in cavities within trabecular bone (RM) is taken to be 1500 g (*ICRP Publication 23*).

7.2.1. *Estimates of absorbed fractions in skeletal tissues*

7.2.1.1. *Photon emitters.* For photon emitters, S is any organ of the body containing the radionuclide and T either the bone surface cells or active red bone marrow. Values of $AF(T \leftarrow S)_i/M_T$, the specific absorbed fraction, are given in *ICRP Publication 23* for the target tissues skeleton and red bone marrow and for all source organs and tissues of interest in this report. The values of specific absorbed fraction given there for skeleton have been taken here as appropriate for bone surface cells.

7.2.1.2. *α emitters uniformly distributed throughout the volume of mineral bone.* In the following discussion it will be assumed that the radionuclides under consideration are uniformly distributed both in cortical and trabecular bone but it will not be assumed that the concentrations in the two sorts of bone are necessarily equal.

(a) $AF(BS \leftarrow \text{TRABECULAR BONE})_i$. Because of the short range of α particles in tissue, surfaces of trabecular bone can be considered essentially infinite and flat for the purposes of α-particle dosimetry. Under this assumption, Thorne (1977) has calculated the fractional energy absorption in a layer 10 μm thick for various depths of burial of α emitters of various energies. The results of these calculations can be used to estimate values of $AF(BS \leftarrow \text{TRABECULAR BONE})_i$, for α emitters of different energies uniformly distributed throughout the volume of trabecular bone. These values range from 0.018 for radionuclides emitting 3 MeV α particles to 0.032 for radionuclides emitting 8 MeV α particles and are in good agreement with those which can be derived from the work of Mays and Sears (1962). In this report a value of 0.025 has been used for all α-emitting radionuclides assumed to be uniformly distributed throughout the volume of trabecular bone.

(b) $AF(BS \leftarrow \text{CORTICAL BONE})_i$. Part of the 120 g of tissue on bone surfaces over which committed dose equivalent is to be estimated lies on the surfaces of Haversian canals. Because of the small dimensions of these canals (~50 μm diameter) some "cross-fire" is likely to occur even with α-particles, thus increasing the dose to surface cells. Even so, $AF(BS \leftarrow \text{CORTICAL BONE})_i$ will be less than the corresponding quantity for trabecular bone as source organ because a larger proportion of α-particle emissions will be out of range of the bone surfaces. Since the bone surface areas of the two types of bone may be taken to be equal (*ICRP Publication 23*) it is readily shown that, in the absence of cross-fire

$$AF(BS \leftarrow CORTICAL\ BONE)_i = \frac{M_x}{M_z} AF(BS \leftarrow TRABECULAR\ BONE)_i$$

where $M_x = 1000$ g is the mass of trabecular bone and $M_z = 4000$ g is the mass of cortical bone (*ICRP Publication 23*);

$$AF(BS \leftarrow TRABECULAR\ BONE)_i = 0.025 \text{ as estimated in 7.2.1.2.(a).}$$

Therefore, in practice, when the effects of cross-fire are included

$$AF(BS \leftarrow CORTICAL\ BONE)_i \geq 0.006.$$

However, if the radionuclide is uniformly distributed throughout cortical bone, AF(BS ← CORTICAL BONE)$_i$ cannot exceed kM_{EZ}/M_Z. Where k is the ratio of the mass stopping power of soft tissue to the mass stopping power of mineral bone for α particles, M_{EZ} is the mass of the surface layer associated with cortical bone and M_Z is the mass of cortical bone as before. Thus, taking M_Z to be 4000 g(*ICRP Publication 23*), M_{EZ} to be 60 g and k to be 1.3 (Thorne, 1976) we obtain

$$0.006 \leq AF(BS \leftarrow CORTICAL\ BONE)_i \leq 0.020$$

Explicit calculations by Spiers (1974a) for ^{226}Ra lead to a value for AF(BS ← CORTICAL BONE)$_i$ of 0.01 and this value has been adopted here for all α-emitting radionuclides distributed throughout the volume of bone.

(c) AF(RM ← TRABECULAR BONE)$_i$. From the data given by Thorne (1977), AF(RM ← TRABECULAR BONE)$_i$ is calculated to be 0.02 for radionuclides emitting 3 MeV α particles and 0.09 for radionuclides emitting 8 MeV α particles, in good agreement with the values that can be derived from the work of Mays and Sears (1962). In this report AF(RM ← TRABECULAR BONE)$_i$ has been taken to be 0.05 for all α-emitting radionuclides assumed to be uniformly distributed throughout the volume of trabecular bone.

(d) AF(RM ← CORTICAL BONE)$_i$. Since active red bone marrow has been defined to lie entirely within the marrow cavities of trabecular bone (*ICRP Publication 23*), AF(RM ← CORTICAL

BONE)$_i$ is considerably smaller than AF(RM $\rightarrow$ TRABECULAR BONE)$_i$ for α-emitting radionuclides uniformly distributed throughout the volume of bone. The work of Whitwell and Spiers (1976) demonstrates that for radionuclides emitting low energy β particles AF(RM $\leftarrow$ CORTICAL BONE)$_i$ is negligible in comparison with AF(RM $\leftarrow$ TRABECULAR BONE)$_i$. Since these β particles have ranges comparable with α particles, this result will also hold true for α-emitting radionuclides. For this reason, AF(RM $\leftarrow$ CORTICAL BONE)$_i$ has been taken to be zero for all α-emitting radionuclides assumed to be uniformly distributed throughout the volume of cortical bone.

7.2.1.3. *α emitters assumed to be on bone surfaces.* In this report radionuclides assumed to be on bone surfaces are taken to be uniformly spread in an infinitely thin layer over the relevant surfaces of bone. This assumption will result in an over-estimate of the true committed dose equivalents received by bone surface cells and active bone marrow because it disregards burial of radioactive deposits by the deposition of new bone mineral. Although such effects might be of considerable importance for several of the radionuclides considered in this report, it was decided that the problem was too complex and that there were insufficient data available to make any reasonable estimate of the reduction in committed dose equivalents due to this cause. This matter will, however, be kept under review, meanwhile it should be noted that values of ALI for the relevant radionuclides may be unduly restrictive.

(a) AF(BS $\leftarrow$ TRABECULAR BONE)$_i$. Thorne (1977) has calculated the fractional energy deposited in a layer 10 μm thick from α-emitting radionuclides distributed as an infinitely thin layer on the surfaces of mineral bone. These calculations show that AF(BS $\leftarrow$ TRABECULAR BONE)$_i$ is 0.43 for radionuclides emitting 3 MeV α particles and 0.15 for radionuclides emitting 8 MeV α particles. These values are in good agreement with those which can be derived from the work of Mays and Sears (1962). In this report a nominal value of 0.25 has been adopted for all α-emitting radionuclides assumed to be on bone surfaces.

(b) AF(BS $\leftarrow$ CORTICAL BONE)$_i$. If cross-fire in small bone cavities can be neglected, AF(BS $\leftarrow$ CORTICAL BONE)$_i$ will be equal to AF(BS $\leftarrow$ TRABECULAR BONE)$_i$ for α-emitting radionuclides distributed in a thin layer on bone surfaces. Monte-Carlo calculations suggest that such cross-fire effects are only important for cavities having radii of curvature of less than 25 μm (Thorne, 1977) and such cavities are uncommon even in cortical bone (Whitwell and Spiers, 1976). For this reason,

AF(BS ← CORTICAL BONE)$_i$ has been taken to be the same as the corresponding value for trabecular bone, i.e. 0.25 for all α-emitting radionuclides assumed to be distributed on bone surfaces.

(c) AF(RM ← TRABECULAR BONE)$_i$. The surfaces of the trabeculae may, to a first approximation, be regarded as infinite flat planes compared with the range of α particles in tissue. If this is assumed and the surfaces are also assumed to be sufficiently far apart to make the effects of cross-fire negligible, then, by simple geometric arguments, AF(RM ← TRABECULAR BONE)$_i$ is equal to 0.5. This value has been adopted in this report.

(d) AF(RM ← CORTICAL BONE)$_i$. Since active red bone marrow has been defined to lie entirely within the trabecular marrow spaces, AF(RM ← CORTICAL BONE)$_i$ will be much less than AF(RM ← TRABECULAR BONE)$_i$ and may, for the purposes of radiological protection, be taken as zero for all α-emitting radionuclides assumed to be on bone surfaces (see section 7.2.1.2).

7.2.1.4. *β emitters uniformly distributed throughout the volume of bone.* Spiers and his coworkers (Spiers, 1968; 1969, 1974a, b; Whitwell and Spiers, 1976) have described how to calculate the average dose rate to cells within 10 μm of bone surfaces and to active red bone marrow from β-emitting radionuclides uniformly distributed throughout the volume of bone. In their nomenclature, D_o is the absorbed dose rate to a small tissue-filled cavity in an infinite extent of mineral bone uniformly contaminated at a level of 1 unit of activity per unit of mass and $D_1 - D_4$ are the absorbed dose rates for the various source and target organs given in Table 7.1 when mineral bone in an adult man is contaminated at a level of 1 unit activity per unit of mass.

Quantities $P_1 - P_4$ are defined such that $P_j = D_j / D_o$, where $j = 1$ to 4. These quantities, which are independent of the level to which the bone is contaminated, are given in Table 7.2 for various radionuclides emitting β particles of average energy $\bar{E}_\beta$ MeV.

Since P_j is the ratio of the dose rate in the target tissue to the dose rate in a small tissue-filled cavity in an infinite extent of mineral bone

Table 7.1. Source and target organs of the skeleton.

Target	Source: Trabecular bone	Source: Cortical bone
Cells near bone surfaces	D_1	D_2
Active red bone marrow	D_3	D_4

Table 7.2. Values of the absorbed dose ratio, P_j, for bone surfaces and red bone marrow from β-emitting radionuclides uniformly distributed throughout the volume of mineral bone. Estimates from Spiers (1974a) and Whitwell and Spiers (1976).

Nuclide	$\bar{E}_\beta$ MeV	P_1	P_2	P_3	P_4
^{45}Ca	0.08	0.21	0.42	0.11	0.00
^{90}Sr	0.20	0.21	0.47	0.22	0.02
^{89}Sr	0.55	0.19	0.47	0.26	0.04
^{90}Y*	0.93	0.16	0.45	0.27	0.06

* Assumed to be uniformly distributed throughout the volume of mineral bone when it is produced in bone from its parent ^{90}Sr.

contaminated to the same level, it may be shown that

$$P_j = \frac{Y_i E_i \mathrm{AF}(T \leftarrow S)_i Q_i C_S}{M_T} \bigg/ \frac{Y_i E_i Q_i C_S k}{M_S}$$

$$= \frac{M_S \mathrm{AF}(T \leftarrow S)_i}{M_T k}$$

and, therefore, that

$$\mathrm{AF}(T \leftarrow S)_i = \frac{k M_T P_j}{M_S} \tag{7.3}$$

where Y_i, E_i, $\mathrm{AF}(T \leftarrow S)_i$ and Q_i are as defined previously;

M_S is the mass of the source organ;
M_T is the mass of the target tissue;
C_S is the activity in the source organ, and
k is the ratio of the stopping power of electrons in soft tissue to the stopping power of electrons in bone.

Thus

$$\mathrm{AF(BS \leftarrow TRABECULAR\ BONE)}_i = \frac{0.12\mathrm{k}}{1} P_1$$

$$\mathrm{AF(BS \leftarrow CORTICAL\ BONE)}_i = \frac{0.12\mathrm{k}}{4} P_2$$

$$\mathrm{AF(RM \leftarrow TRABECULAR\ BONE)}_i = \frac{1.5\mathrm{k}}{1} P_3$$

$$\mathrm{AF(RM \leftarrow CORTICAL\ BONE)}_i = \frac{1.5\mathrm{k}}{4} P_4$$

where 1.5 kg is the mass of active red bone marrow;
4 kg is the mass of cortical bone;

1 kg is the mass of trabecular bone (*ICRP Publication 23*);
0.12 kg is the mass of the layer between 0 and 10 μm from the surfaces of cortical and trabecular bone, and
k is taken to be 1.07, a value appropriate for electrons over a wide range of energies (Berger and Seltzer, 1966).

Values of the absorbed fractions for a number of radionuclides are given in Table 7.3.

Since the main contribution to dose equivalent in the red bone marrow arises from activity in trabecular bone, it is reasonable to neglect the contribution to dose equivalent in red bone marrow from β activity in cortical bone and to take nominal values of absorbed fractions for β-emitting radionuclides uniformly distributed throughout the volume of mineral bone. Hence the values in Table 7.3 may be reduced to the following representative values

$$AF(BS \leftarrow TRABECULAR\ BONE)_i = 0.025$$

$$AF(BS \leftarrow CORTICAL\ BONE)_i = 0.015$$

$$AF(RM \leftarrow TRABECULAR\ BONE)_i = 0.35$$

$$AF(RM \leftarrow CORTICAL\ BONE)_i = 0.0.$$

7.2.1.5. *β emitters assumed to be on bone surfaces.*

(a) AF(BS ← TRABECULAR BONE)$_i$. If $\bar{E}_\beta$ is less than 0.05 MeV, then the average range of the β particles will be similar to the range of α particles in tissue and AF(BS → TRABECULAR BONE)$_i$ will be about 0.25. However, if $\bar{E}_\beta$ is about 1 MeV the average range of the β particles will be many times the average linear dimensions of a marrow cavity or trabeculum and AF(BS ← TRABECULAR BONE)$_i$ may be taken as 0.025, the same as for β emitters distributed uniformly throught the volume of

Table 7.3. Values of absorbed fractions for β emitters uniformly distributed throughout the volume of mineral bone

Nuclide	^{45}Ca	^{90}Sr	^{89}Sr	^{90}Y[a]
$\bar{E}_\beta$	0.08	0.20	0.55	0.93
AF (BS ← TRABECULAR BONE)$_i$	0.027	0.027	0.024	0.021
AF (BS ← CORTICAL BONE)$_i$	0.013	0.015	0.015	0.014
AF (RM ← TRABECULAR BONE)$_i$	0.18	0.35	0.42	0.43
AF (RM ← CORTICAL BONE)$_i$	0.000	0.007	0.014	0.021

[a] Assumed to be uniformly distributed throughout the volume of mineral bone when it is produced in bone from its parent ^{90}Sr.

bone. There is, therefore, a considerable difference between the absorbed fractions for low and high energy β emitters on the surface of bone. It is impossible to assign a nominal value to AF(BS $\leftarrow$ TRABECULAR BONE)$_i$ which would be appropriate over this entire range of energies. Two nominal values have, therefore, been chosen;

(i) AF(BS $\leftarrow$ TRABECULAR BONE)$_i$ = 0.25, for all values of $\bar{E}_\beta$ less than 0.2 MeV, and

(ii) AF(BS $\leftarrow$ TRABECULAR BONE)$_i$ = 0.025, for all values of $\bar{E}_\beta$ greater than or equal to 0.2 MeV.

The dividing line of 0.2 MeV between the two classes has been chosen from a consideration of the absorbed dose rate to surface cells from activity on the surface below those cells and the absorbed dose rate to these cells from activity on other areas of trabecular bone. The first of these two components falls rapidly as $\bar{E}_\beta$ increases, while the second rises rather more slowly with increasing $\bar{E}_\beta$. Since the value of AF(BS $\leftarrow$ TRABECULAR BONE)$_i$ is not much larger for $\bar{E}_\beta$ = 0.2 MeV than it is for $\bar{E}_\beta$ = 1 MeV but increases markedly with decreasing energy below 0.2 MeV (Spiers, 1974a) the former energy is chosen to mark the boundary between the two classes.

(b) AF(BS $\leftarrow$ CORTICAL BONE)$_i$. By arguments similar to those above, two nominal values of AF(BS $\leftarrow$ CORTICAL BONE)$_i$ have also been adopted:

(i) AF(BS $\leftarrow$ CORTICAL BONE)$_i$ = 0.25 for $\bar{E}_\beta$ less than 0.2 MeV;

(ii) AF(BS $\leftarrow$ CORTICAL BONE)$_i$ = 0.015 for $\bar{E}_\beta$ greater than or equal to 0.2 MeV.

(c) AF(RM $\leftarrow$ TRABECULAR BONE)$_i$. For $\bar{E}_\beta$ less than 0.05 MeV the average range of the β particles will be similar to the range of α particles in tissues. Thus, in this case, AF(RM $\leftarrow$ TRABECULAR BONE)$_i$ may be taken as 0.5, the value used for α emitters assumed to be deposited on bone surfaces. If, however, $\bar{E}_\beta$ is about 1 MeV then the range of the β particles will be considerably greater than the average thickness of trabecular bone. In this case, AF(RM $\leftarrow$ TRABECULAR BONE)$_i$ will be about 0.4, the value appropriate to β emitters of this energy uniformly distributed throughout the volume of trabecular bone (Table 7.3). It is reasonable, therefore, to take AF(RM $\leftarrow$ TRABECULAR BONE)$_i$ to be 0.5 for all β emitters mainly distributed on bone surfaces.

(d) AF(RM ← CORTICAL BONE)$_i$. Since active red bone marrow is assumed to be entirely contained within trabecular bone AF(RM ← CORTICAL BONE)$_i$ is much less than AF(RM ← TRABECULAR BONE)$_i$ for $\bar{E}_\beta \lessapprox 0.05$ *MeV* (see Section 7.2.1.2). If, however, $\bar{E}_\beta$ is about 1 MeV then the range of the β particles will be considerably greater than the average thickness of cortical bone. In this case AF(RM ← CORTICAL BONE)$_i$ will be about 0.02, the value appropriate to β emitters of this energy uniformly distributed throughout the volume of cortical bone, as given in Table 7.3. Thus, again AF(RM ← CORTICAL BONE)$_i$ is much less than AF(RM ← TRABECULAR BONE)$_i$ and for the purposes of radiological protection may be taken to be zero.

7.2.1.6. *Fission fragments and recoil atoms.* Fission fragments have ranges in tissue of about 20 μm (Green *et al.*, 1977) which is about the same as that for 3 MeV α particles. Values of absorbed fractions for fission fragments are taken to be the same as for α particles in all cases. The energies of recoil atoms are negligible in comparison with the primary emissions and are disregarded for purposes of bone dosimetry.

7.2.1.7. *Summary of values of absorbed fractions.*

7.2.2. *Estimates of the number of transformations in trabecular and cortical bone*

7.2.2.1. *Parent radionuclides.* With the present state of knowledge it is not possible to estimate $U_{\text{TRABECULAR BONE}}$ and $U_{\text{CORTICAL BONE}}$ independently for most bone-seeking radionuclides. Only for the alkaline earth elements which are incorporated into mineral bone is there a comprehensive model for the distribution of activity between cortical and trabecular bone (*ICRP Publication 20*). Although $U_{\text{TRABECULAR BONE}}$ and $U_{\text{CORTICAL BONE}}$ cannot be calculated with any precision for most bone-seeking radionuclides, nevertheless estimates of these quantities can be made.

(a) If a radionuclide is deposited uniformly on all bone surfaces and is removed at the same rate from both cortical and trabecular bone, then, since the surface areas of cortical and trabecular bone are assumed to be equal, $U_{\text{TRABECULAR BONE}} = U_{\text{CORTICAL BONE}} = 0.5\ U_{\text{MINERAL BONE}}$.

(b) If a radionuclide redistributes, in a time short compared with

Table 7.4. Recommended absorbed fractions for dosimetry of radionuclides in bone

Source organ	Target organ	Class of radionuclide (see Section 7.2)				
		α emitter uniform in volume	α emitter on bone surfaces	β emitter uniform in volume	β emitter on bone surfaces $\bar{E}_\beta \geq 0.2$ MeV	β emitter on bone surfaces $\bar{E}_\beta < 0.2$ MeV
Trabecular bone	Bone surfaces (BS)	0.025	0.25	0.025	0.025	0.25
Cortical bone	Bone surfaces (BS)	0.01	0.25	0.015	0.015	0.25
Trabecular bone	Red bone marrow (RM)	0.05	0.5	0.35	0.5	0.5
Cortical bone	Red bone marrow (RM)	0.0	0.0	0.0	0.0	0.0

both its radioactive half-life and its time of residence in bone, such that it uniformly contaminates all bone mass, then;

$$\frac{U_{\text{TRABECULAR BONE}}}{U_{\text{CORTICAL BONE}}} \simeq \frac{M_x}{M_z}$$

where M_x is the mass of trabecular bone and M_z is the mass of cortical bone, taken to be 1 and 4 kg, respectively,
from which it follows that $U_{\text{TRABECULAR BONE}} \simeq 0.2\ U_{\text{MINERAL BONE}}$ and that $U_{\text{CORTICAL BONE}} \simeq 0.8\ U_{\text{MINERAL BONE}}$. Therefore, when specific information on the distribution of a radionuclide is not available, the following approximation is used

(i) $U_{\text{TRABECULAR BONE}} = U_{\text{CORTICAL BONE}} = 0.5\ U_{\text{MINERAL BONE}}$ for radionuclides assumed to be on bone surfaces,

(ii) $U_{\text{TRABECULAR BONE}} = 0.2\ U_{\text{MINERAL BONE}}$ and $U_{\text{CORTICAL BONE}} = 0.8\ U_{\text{MINERAL BONE}}$ for radionuclides assumed to be uniformly distributed throughout the volume of mineral bone.

For radioisotopes of the alkaline earths, values of U for cortical and trabecular bone can be obtained directly from the retention functions given in *ICRP Publication 20.*

APPENDIX C

Summary of Alkaline Earth Model Given in *ICRP* Publication 20, *Alkaline Earth Metabolism in Adult Man*[12]

Retention in bone volume is described by two functions, one for compact bone and one for cancellous bone. They are, respectively:

$$R_{COMPVOLUME} = p\beta\epsilon^{b}(t + \theta)^{-b}e^{-r\lambda t}$$

$$R_{CANVOLUME} = p(1 - \beta)\epsilon^{b}(t + \theta)^{-b}e^{-\sigma r\lambda t}$$

Theta (θ) is a time-dependent factor which is defined in terms of the whole body retention function, R:

$$\theta = [f_C 0.8c\lambda\omega(1 - R)/p\epsilon^{b}\beta\eta k]^{-1/b}$$

$$R = (1 - p)e^{-mt} + p\epsilon^{b}(t + \epsilon)^{-b}[\beta e^{-r\lambda t} + (1 - B)e^{-\sigma r\lambda t}]$$

Total bone surface retention is defined as the solution to the following differential equation:

$$dR_{SURFACE}/dT = \lambda_{SUR}(c_{SURFACE}S - R_{SURFACE})$$

Where S is the fraction of injected activity per gram calcium in the blood plasma; S is related to the whole body retention function by $S = (-1/\eta k)(dR/dt)$.

Retention on cancellous bone surface = $\frac{1}{2}R_{SURFACE}$ and retention on compact bone surfaces = $\frac{1}{2}R_{SURFACE}$.

The retention functions exclude the effect of radioactive decay and assume that a single intravenous injection has been made. The param-

[12] ICRP, 1973.

eters take on different values for Ca, Sr, Ba, and Ra which are given in Table 27 of ICRP Publication 20 (ICRP, 1973).

New values of some of the parameters for Ra have recently been determined by Schlenker *et al.* (1982) on the basis of new data and should be used in place of those originally published.

Definitions of the parameters in the foregoing equations are:

b—power function slope (related to diminution and diffusion)

β—fraction of bone volume activity deposited in compact bone

c—whole body calcium (also used for total bone calcium)

$c_{SURFACE}$—calcium content of bone surface pool

ϵ—power function time shift (related to the turnover time of an initial pool)

η—excretory discrimination factor for the given alkaline earth relative to calcium

f_C—ratio of the total activity deposited in compact bone volume to that deposited in new compact bone

k—rate of endogenous calcium excretion (urinary plus fecal)

λ—apposition rate and resorption rate of compact bone

λ_{SUR}—rate constant of bone surface pool

p—early exponential coefficient

r—recycling correction

σ—ratio of apposition-resorption rate of cancellous to that of compact bone

ω—discrimination factor for alkaline earth relative to calcium from blood to new bone.

APPENDIX D

NCRP Scientific Committee 57 Task Groups

The following lists represents the membership of task groups during the period of preparation of the Phase I Report. Some task groups are no longer active and some have an altered membership for Phase II activities.

Task Group 1—Concepts and Terminology
R. Thompson

Task Group 2—Respiratory Tract Models
D. Craig, Chairman

J. Brain
R. Cuddihy
G. Kanapilly (Deceased)
R. Phalen
B. Stuart
D. Swift

Task Group 3—General Metabolic Models
B. Boecker, Chairman

S. Cohn
J. Poston
J. Rundo
M. Wrenn

Task Group 4—Radon and Daughters
N. Harley, Chairman

F. Cross
V. Archer (Advisor)
J. Harley (Consultant)
B. Stuart
D. Morken (Advisor)

Task Group 5—Gastrointestinal Tract Models
M. Sullivan, Chairman

F. Cross
C. Lushbaugh

Task Group 6—Bone Problems
M. Goldman, Chairman

J. Arnold
W. Jee
T. Jones
R. Rowland
B. Schlenker
D. Weber
B. Stover
R. Evans (Consultant)

Task Group 7—Thyroid Cancer Risk
H. Maxon, Chairman

D. Becker
S. Book
B. Edwards
S. Thomas
E. Saenger (Advisor)

Task Group 8—Leukemia Risk
S. Marks, Chairman

E. Cronkite
T. Jones
B. Greenburg
K. Stitzel (Consultant)
L. Rosenblatt
H. Ragan
F. Wilson

Task Group 9—Lung Cancer Risk
F. Hahn, Chairman

V. Archer
C. Sanders

Task Group 10—Liver Cancer Risk
G. Taylor, Chairman

A. Brooks
A. Brues
C. Mays
N. Telles

Task Group 11—Genetic Risk
A. Brooks, Chairman

D. Grahn
P. Selby
W. Russell

Task Group 12—Strontium
R. D. Lloyd, Chairman

H. Andrews
P. Durbin
H. Ragan
R. Poole
R. Jones
J. Parks

APPENDIX E

Glossary

absorbed dose: The quotient of $d\bar{\epsilon}$ by dm where $d\bar{\epsilon}$ is the mean energy imparted by ionizing radiation to the matter in a volume element and dm is the mass of the matter in that volume element, i.e., the absorbed dose, $D = d\bar{\epsilon}/dm$. The special unit of absorbed dose is the rad. In the SI it is the gray (Gy).

absorbed fraction (AF): The fraction of energy emitted as a specified radiation type in a specified source tissue which is absorbed in a specified target tissue.

activity median aerodynamic diameter (AMAD): The diameter of a unit density sphere with the same terminal settling velocity in air as that of the aerosol particle whose activity is the median for the entire aerosol.

anatomic models: Stylized descriptions of the location, size and shape of selected portions of the anatomy of man, woman or child.

annual limit on intake (ALI): The activity of a radionuclide that taken alone would irradiate a person, represented by Reference Man, to the limit set for occupational exposure by recommending and regulating bodies.

committed dose equivalent (H_{50}): The total dose equivalent averaged throughout any tissue in the 50 years after intake of a radionuclide into the body.

committed effective dose equivalent: The weighted sum of organ committed dose equivalents using the ICRP organ dose weighting factors.

cortical bone: Equivalent to "Compact Bone" in ICRP Publication 20, i.e., any bone with a surface/volume ratio less than 60 $cm^2\ cm^{-3}$; in Reference Man the total mass of cortical bone is 4000 g.

critical organ: The organ determining the maximum permissible concentrations in air or water. Theoretically it was the organ suffering the most damage or whose functions were most essential, but in practice it was usually the organ with the highest concentration of radionuclide.

derived air concentration (DAC): Equals the ALI of a radionuclide divided by the volume of air inhaled by Reference Man in a working year (i.e., 2.4×10^3 m^3). The unit of DAC is Bq m^{-3}.

derived limits: Any of a number of limits such as "Derived Air Concentration" (DAC) calculated from the basic radiation protection limits by application of standard parameters for the pertinent biological system.

dose commitment: An early term for expression of the legacy of dose from a single intake of a radionuclide. No specific time interval is designated in contrast to "committed dose equivalent."

dose distribution factor: A factor used to modify the absorbed dose when distribution of the dose is not nominally uniform. It has usually been applied to gross rather than microscopic distributions.

dose equivalent (*H*): A quantity used for radiation protection purposes that expresses on a common scale for all radiations, the irradiation incurred by exposed persons. It is defined as the product of the absorbed dose and certain modifying factors. The unit of dose equivalent is currently the rem (see rem) or the sievert (see Sv) (NCRP, 1985d).

dose equivalent commitment: The infinite-time integral of the *per caput* dose equivalent rate in any given organ or tissue for a specified population.

dose rate effectiveness factor (DREF): A factor used to account for the difference in effectiveness of low-LET radiations at low versus high dose rates. In NCRP Report No. 64 it is quantitated as: "The factor by which linear interpolation from data obtained at high doses and dose rates overestimates the risk per unit absorbed dose of radiation delivered at very low doses and/or dose rates. Using the $\alpha D + \beta D^2$ formulation, at a given dose it is $(\alpha D + \beta D^2/\alpha D)$."

effective absorbed energy: A term, not now used, for the product of the energy released by a radionuclide, the relative biological effectiveness of the type of radiation emitted and other modifying factors such as the fraction of disintegrations having a given energy, the ratio of daughter atoms to parent atoms, the *n* factor (which see), etc.

effective dose equivalent (H_E): The product of the dose equivalent in a tissue (T) and the weighting factor representing its proportion of the stochastic risk resulting from irradiation of tissue (T) to the total risk when the whole body is irradiated uniformly.

gastrointestinal tract model: Model used to stylize the behavior of radionuclides in the gastrointestinal tract of man.

gray (Gy): The unit of absorbed dose in the SI (NCRP, 1985d). 1 Gy $= 1$ J $kg^{-1} = 100$ rad.

hot particle: An aggregation of radionuclide atoms.

hot spot: An area of unusually high concentration of a radionuclide in an organ or tissue, frequently on a microscopic scale.

internal emitter: A term used for a radionuclide deposited in the body.

maximum permissible concentration (MPC): The concentration of air (MPC_a) or water (MPC_w) that would lead at equilibrium to an amount of radionuclide in the critical organ that would just deliver the maximum permissible dose rate to that organ. These were calculated for either 40 hrs. per week or 168 hrs. per week intakes.

metabolic contants: Factors for the rate of transfer of radionuclides at various steps in a metabolic cycle.

metabolic model: A mathematical description of the behavior in the metabolic processes of cells, tissues, organs and organisms. It is most frequently used to describe distribution among tissues and excretion.

MIRD: Medical Internal Radiation Dose. A committee of the Society of Nuclear Medicine.

mucociliary pathway: Those portions of the respiratory tract lined with cilia that propel materials to the mouth.

***n*:** A factor, termed at various times the "non-uniform distribution factor" or "the relative hazard factor," used to adjust the absorbed dose in bone for observed differences in the biological effectiveness of different radionuclides not dependent on radiation quality.

non-stochastic effects: Effects for which the severity of the effect varies with the dose, and for which a threshold may therefore occur.

nuclide: A species of atom characterized by the constitution of its nucleus.

organ weighting factors: Factors indicating the proportion of the total risk to the body of cancer and heredity defects attributable to a given organ or group or organs (i.e., organ dose weighting factors).

osteogenic potential: Potential to produce bone cancer, osteosarcoma.

permissible organ burden: The quality of a radionuclide in any organ that just delivers the maximum permissible dose rate.

quality factor (Q): A factor chosen to encompass appropriate values of the relative biological effectiveness (RBE), but to be independent of the organ and tissue or of the biological endpoint under consideration. It is also chosen to be a smooth function of the collision stopping power of the charged particles (ICRU, 1980).

rad: A unit of absorbed dose: One rad is 0.01 joules absorbed per kilogram of any material. (Also defined as 100 ergs per gram.) Being replaced by the gray (NCRP, 1985d).

radionuclide: A nuclide that is radioactive.

radionuclide exposure: The situation leading to intake of a radionuclide and/or the situation after a radionuclide has been deposited in an organ or tissue.

RBE dose: An old term with the same meaning as dose equivalent now, which see.

Reference Man: A person with the anatomical and physiological characteristics defined in the report of the ICRP Task Group on Reference Man (ICRP Publication 23).

relative biological effectiveness (RBE): A ratio of the absorbed dose of a reference radiation to the absorbed dose of a test radiation to produce the same level of biological effect, other conditions being equal (ICRU, 1979).

rem: A unit of dose equivalent. The product of the absorbed dose in rad and modifying factors. Being replaced by the sievert (NCRP, 1985d).

respiratory tract model: A general term for models of the behavior of particles and gases in the human respiratory tract. See also task group Lung model.

sievert (Sv): The measure of dose equivalent in the SI. It is the product of absorbed dose in grays and any modifying factors. It is equivalent to 100 rem.

specific effective energy (SEE $(T \leftarrow S)_i$): The energy (MeV), suitably modified for quality factor, imparted per gram of a target tissue (T) as a consequence of the emission of a specified radiation (i) from a transformation occurring in source tissue (S).

stochastic effects: Defined by ICRP as malignant or hereditary disease for which the probability of an effect occurring, rather than its severity, is a function of dose without threshold. More generally stochastic means random in nature.

stroma: Supporting framework of an animal organ.

surface-seeking radionuclides: Radionuclides that both deposit on and remain for a considerable period on the surface of bone structure. To be contrasted with "Volume-seekers" that exchange for bone mineral.

task group lung model: The model for behavior of particles in the respiratory tract of man developed by the Task Group on Lung Dynamics of the ICRP.

trabecular bone: Equivalent to "Cancellous Bone" in ICRP Publication 20, i.e., any bone with a surface/volume ratio greater than 60 $cm^2\ cm^{-3}$; in Reference Man trabecular bone has a mass of 1000 g.

transformation: The process of radioactive decay in which one radionuclide is transformed into another.

References

BARON, R., VIGNERY, A., AND TRANVAN, P. (1982). "The significance of lacunar erosion without osteoclasts: Studies on the reversal phase of the remodeling sequence," page 35 in *Bone Histomorphometry*, Jee, W.S. S. and Parfitt, A.M., Eds. (Societe Nouvelle de Publications Medicales et Dentaires, Paris).

BERGER, M. J. (1974). "Beta-ray doses in tissue-equivalent material immersed in a radioactive cloud," Health Phys. **26**, 1.

DOLPHIN, G. W. (1974). "Hot particles," Radiol. Prot. Bull. **8**, 8.

DOLPHIN, G. W. (1975). "The hot particle problem," Radiol. Prot. Bull. **12**, 18.

DOLPHIN, G. W. AND EVE, I. S. (1966). "Dosimetry of the gastrointestinal tract," Health Phys. **12**, 163.

DUNNING, D. E., JR., BERNARD, S. R., WALSH, P. J., KILLOUGH, G. G., AND PLEASANT, J. C. (1979). *Estimates of Internal Dose Equivalent for 22 Target Organs for Radionuclides Occurring in Routine Releases from Nuclear Fuel-Cycle Facilities*, Vol. 2, Report NUREG/CR-0150, ORNL/NUREG/TM-0190 (Oak Ridge National Laboratory, Oak Ridge, Tennessee).

EVE, I. S. (1966). "A Review of the physiology of the gastrointestinal tract in relation to radiation dose from radioactive materials," Health Phys. **12**, 131.

GMI (1978). German Ministry of the Interior (Bonn), "On the toxicity of inhaled particles with special reference to plutonium," Radiat. Environ. Biophys. **15**, 1.

HARRISON, G. E. (1981). "Whole body retention of the alkaline earths in adult man," Health Phys. **40**, 95.

HEALY, J. W. (1981). "Commitment, equivalent, and collective dose," Health Phys. **41**, 379.

HOENES, G. R., AND SOLDAT, J. K. (1977). *Age-specific Radiation Dose Commitment Factors for a One-year Chronic Intake*. Report NUREG-0172 (U.S. Nuclear Regulatory Commission, Washington, D.C.).

ICRP (1959). International Commission on Radiological Protection, *Report of Committee II on Permissible Dose for Internal Radiation*, ICRP Publication 2 (Pergamon Press, New York, New York).

ICRP (1964). International Commission on Radiological Protection, *Recommendations as Amended 1959 and Revised 1962*. ICRP Publication 6 (Pergamon Press, New York, New York).

ICRP (1966a). International Commission on Radiological Protection, *Recommendations of the ICRP* (adopted September 17, 1965), ICRP Publication 9 (Pergamon Press, New York, New York).

ICRP (1966b). International Commission on Radiological Protection, Committee II, Task Group on Lung Dynamics, "Deposition and retention models for internal dosimetry of the human respiratory tract," Health Phys. **12**, 173.

ICRP (1968a) International Commission on Radiological Protection, *Report of Committee IV on Evaluation of Radiation Doses to Body Tissues from Internal Contamination Due to Occupational Exposure*, ICRP Publication 10 (Pergamon Press, New York, New York).

ICRP (1968b). International Commission on Radiological Protection, *A Review of the Radiosensitivity of the Tissues of Bone*, ICRP Publication 11 (Pergamon Press, New York, New York).

ICRP (1971). International Commission on Radiological Protection, *The Assessment of Internal Contamination Resulting from Recurrent or Prolonged Uptakes*, ICRP Publication 10A (Pergamon Press, New York, New York).

ICRP (1972). International Commission on Radiological Protection, *The Metabolism of Compounds of Plutonium and Other Actinides*, ICRP Publication 19 (Pergamon Press, New York, New York).

ICRP (1973). International Commission on Radiological Protection, *Alkaline Earth Metabolism in Adult Man*, ICRP Publication 20 (Pergamon Press, New York, New York).

ICRP (1975). International Commission on Radiological Protection, *Reference Man: Anatomical, Physiological, and Metabolic Characteristics*, ICRP Publication 23 (Pergamon Press, New York, New York).

ICRP (1977). International Commission on Radiological Protection, *Recommendations of the International Commission on Radiological Protection*, ICRP Publication 26 (Pergamon Press, New York, New York).

ICRP (1978). International Commission on Radiological Protection, "Recommendations of 1978 Stockholm Meeting," Phys. Med. Biol. **23**, 1209.

ICRP (1979a). International Commission on Radiological Protection, *Limits for Intakes of Radionuclides by Workers*, ICRP Publication 30, Part 1, Annals of the ICRP 2, Nos. 3–4 (Pergamon Press, New York, New York).

ICRP (1979b). International Commission on Radiological Protection, ICRP Publication 30, Supplement of Part 1, Annals of the ICRP 3, Nos. 1–4 (Pergamon Press, New York, New York).

ICRP (1980a). International Commission on Radiological Protection, ICRP Publication 31, Annals of the ICRP 4, Nos. 1–2 (Pergamon Press, New York, New York).

ICRP (1980b). International Commission on Radiological Protection, ICRP Publication 30, Part 2, Annals of the ICRP 4, Nos. 3–4 (Pergamon Press, New York, New York).

ICRP (1981a). International Commission on Radiological Protection, ICRP Publication 30, Supplement to Part 2, Annals of the ICRP 5, Nos. 1–6 (Pergamon Press, New York, New York).

ICRP (1981b). International Commission on Radiological Protection, ICRP Publication 30, Part 3, Annals of the ICRP 6, Nos. 2–3 (Pergamon Press, New York, New York).

ICRP (1982). International Commission on Radiological Protection, ICRP Publication 30, Supplements A and B to Part 3, Annals of the ICRP 7, 8, Nos. 1–3 (Pergamon Press, New York, New York).

ICRP (1984). International Commission on Radiological Protection, *A Compilation of Major Concepts and Quantities in Use by ICRP*, ICRP Publication 42, Annals of the ICRP 14, No. 4 (Pergamon Press, New York, New York).

ICRP (1985). International Commission on Radiological Protection, "The value of the quality factor in the case of neutrons," Health Phys. *48*, 828.

ICRU (1963). International Commission on Radiation Units and Measurements, *Radiological Dosimetry*, ICRU Report 10e (International Commission on Radiation Units and Measurements, Bethesda, Maryland).

ICRU (1968). International Commission on Radiation Units and Measurements, *Radiation Quantities and Units*, ICRU Report 11 (International Commission on Radiation Units and Measurements, Bethesda, Maryland).

ICRU (1973). International Commission on Radiation Units and Measurements, *Dose Equivalent*, Supplement to ICRU Report 11 (International Commission on Radiation Units and Measurements, Bethesda, Maryland).

ICRU (1979). International Commission on Radiation Units and Measurements, *Quantitative Concepts and Dosimetry in Radiobiology*, ICRU Report 30 (International Commission on Radiation Units and Measurements, Bethesda, Maryland).

ICRU (1980). International Commission on Radiation Units and Measurements, *Radiation Quantities and Units*, ICRU Report 33 (International Commission on Radiation Units and Measurements, Bethesda, Maryland).

JACQUEZ, J. A. (1972). *Compartmental Analysis in Biology and Medicine* (Elsevier, Amsterdam).

JOHNSON, J. R. AND MYERS, R. C. (1981). "Alkaline earth metabolism: A model useful in calculating organ burdens, excretion rates and committed effective dose equivalent conversion factors," Radiat. Prot. Dosim. **1**, 87.

KILLOUGH, G. G., DUNNING, D. E., JR., BERNARD, S. R., AND PLEASANT, J. C. (1978). *Estimates of Internal Dose Equivalent for 22 Target Organs for Radionuclides Occurring in Routine Releases from Nuclear Fuel-Cycle Facilities*, Vol. I. Report NUREG/CR-0150, ORNL/NUREG/TM-0190 (Oak Ridge National Laboratory, Oak Ridge, Tennessee).

KIMMEL, D. B. (1981). "A light microscopic description of osteoprogenitor cells or remodeling bone in the adult," page 181 in: *Bone Histomorphometry*, Jee, W.S.S. and Parfitt, A.M. Eds. (Societe Nouvelle de Publications Medicales et Dentaires, Paris).

LORD, B. I., TESTA, N. G., AND HENDRY, J. H. (1975). "The relative spatial distributions of CFU_S and CFU_C in the normal femur," Blood **46**, 65.

MARCUS, A. H. AND BECKER, A. (1980). "Alkaline earth metabolism: The ICRP model reformulated as a semi-Markov model," Health Phys. **38**, 825.

MIRD (1968, 1969a, 1969b, 1970, 1971). Medical Internal Radiation Dose Committee, Supplements 1, 2, 3, 4, 5 (Pamphlets 1–3, 4, 5, 6, 7, 8) to J. Nucl. Med. (Available from MIRD Pamphlets, Society of Nuclear Medicine, 475 Park Avenue South, New York, New York 10016).

NAS (1976). National Academy of Sciences, National Research Council, *Health Effects of Alpha-Emitting Particles in the Respiratory Tract*, Report to the Environmental Protection Agency, Office of Radiation Programs, EPA 520/4-76-013 (Environmental Protection Agency, Washington, D.C.).

NCRP (1959). National Council on Radiation Protection and Measurements, *Maximum Permissible Body Burdens and Maximum Permissible Concentrations of Radionuclides in Air and in Water for Occupational Exposure*, NCRP Report No. 22 (National Council on Radiation Protection and Measurements, Bethesda, Maryland).

NCRP (1970). National Council on Radiation Protection and Measurements, *Precautions in the Management of Patients Who Have Received Therapeutic Amounts of Radionuclides*, NCRP Report No. 37 (National Council on Radiation Protection and Measurements, Bethesda, Maryland).

NCRP (1971). National Council on Radiation Protection and Measurements, *Basic Radiation Protection Criteria*, NCRP Report No. 39 (National Council on Radiation Protection and Measurements, Bethesda, Maryland).

NCRP (1975a) National Council on Radiation Protection and Measurements, *Review of the Current State of Radiation Protection Philosophy*, NCRP Report No. 43 (National Council on Radiation Protection and Measurements, Bethesda, Maryland).

NCRP (1975b). National Council on Radiation Protection and Measurements, *Krypton-85 in the Atmosphere—Accumulation, Biological Significance, and Control Technology*, NCRP Report No. 44 (National Council on Radiation Protection and Measurements, Bethesda, Maryland).

NCRP (1975c). National Council on Radiation Protection and Measurements, *Alpha-Emitting Particles in Lungs*, NCRP Report No. 46 (National Council on Radiation Protection and Measurements, Bethesda, Maryland).

NCRP (1976). National Council on Radiation Protection and Measurements, *Tritium Measurement Techniques*, NCRP Report No. 47 (National Council on Radiation Protection and Measurements, Bethesda, Maryland).

NCRP (1977a). National Council on Radiation Protection and Measurements, *Cesium-137 From the Environment to Man: Metabolism and Dose*, NCRP Report No. 52 (National Council on Radiation Protection and Measurements, Bethesda, Maryland).

NCRP (1977b). National Council on Radiation Protection and Measurements, *Protection of the Thyroid Gland in the Event of Releases of Radioiodine*, NCRP Report No. 55 (National Council on Radiation Protection and Measurements, Bethesda, Maryland).

NCRP (1978). National Council on Radiation Protection and Measurements, *Physical, Chemical, and Biological Properties of Radiocerium Relevant to Radiation Protection Guidelines*, NCRP Report No. 60 (National Council on Radiation Protection and Measurements, Bethesda, Maryland).

NCRP (1979a). National Council on Radiation Protection and Measurements, *Tritium in the Environment*, NCRP Report No. 62 (National Council on Radiation Protection and Measurements, Bethesda, Maryland).

NCRP (1979b). National Council on Radiation Protection and Measurements,

Tritium and Other Radionuclide Labeled Organic Compounds Incorporated in Genetic Material, NCRP Report No. 63 (National Council on Radiation Protection and Measurements, Bethesda, Maryland).

NCRP (1980a). National Council on Radiation Protection and Measurements, *Influence of Dose and its Distribution in Time on Dose—Effect Relationships for Low Let Radiation,* NCRP Report No. 64 (National Council on Radiation Protection and Measurements, Bethesda, Maryland).

NCRP (1980b). National Council on Radiation Protection and Measurements, *Management of Persons Accidentally Contaminated with Radionuclides,* NCRP Report No. 65 (National Council on Radiation Protection and Measurements, Bethesda, Maryland).

NCRP (1982). National Council on Radiation Protection and Measurements, *Nuclear Medicine-Factors Influencing the Choice and Use of Radionuclides in Diagnosis and Therapy,* NCRP Report No. 70 (National Council on Radiation Protection and Measurement, Bethesda, Maryland).

NCRP (1983). National Council on Radiation Protection and Measurements, *Iodine-129: Evaluation of Releases from Nuclear Power Generation,* NCRP Report No. 75 (National Council on Radiation Protection and Measurements, Bethesda, Maryland).

NCRP (1984a). National Council on Radiation Protection and Measurements, *Exposures from the Uranium Series with Emphasis on Radon and its Daughters,* NCRP Report No. 77 (National Council on Radiation Protection and Measurements, Bethesda, Maryland).

NCRP (1984b). National Council on Radiation Protection and Measurements, *Evaluation of Occupational and Environmental Exposures to Radon and Radon Daughters in the United States,* NCRP Report No. 78 (National Council on Radiation Protection and Measurements, Bethesda, Maryland).

NCRP (1985a). National Council on Radiation Protection and Measurements, *A Handbook of Radioactivity Measurements Procedures,* NCRP Report No. 58, 2nd Ed. (National Council on Radiation Protection and Measurements, Bethesda, Maryland).

NCRP (1985b). National Council on Radiation Protection and Measurements, *Carbon-14 in the Environment,* NCRP Report No. 81 (National Council on Radiation Protection and Measurements, Bethesda, Maryland).

NCRP (1985c). National Council on Radiation Protection and Measurements, *The Experimental Basis for Absorbed Dose-Calculations, in Medical Uses of Radionuclides,* NCRP Report No. 83 (National Council on Radiation Protection and Measurements, Bethesda, Maryland).

NCRP (1985d). National Council on Radiation Protection and Measurements, *SI Units in Radiation Protection and Measurements,* NCRP Report No. 82 (National Council on Radiation Protection and Measurements, Bethesda, Maryland).

POSTON, J. R. AND SNYDER, W. S. (1974). "A model for exposure to a semi-infinite cloud of a photon emitter," Health Phys. **26**, 287.

PRIEST, N. D. AND HUNT, B. W. (1979). "Calculation of annual limits of intake for plutonium-239 in man using a bone model which allows for plutonium burial and recycling," Phys. Med. Biol. **24**, 525.

SCHLENKER, R. A., KEANE, A. T., AND HOLTZMAN, R. B. (1982). "The retention of ^{226}Ra in human soft tissue and bone: implications for the ICRP Publication 20 alkaline earth model," Health Phys. **42**, 671.

SKRABLE, K., FRENCH, C., CHABOT, G., AND MAJOR, A. (1974). "A general equation for the kinetics of linear first order phenomena and suggested applications," Health Phys. **27**, 155.

SNYDER, W. S. (1970). "Estimation of absorbed fraction energy from photon sources in body organs," in *Medical Radionuclides: Radiation Dose and Effects*, Cloutier, R.J., Edwards, C.L., and Snyder, W.S., Eds., AEC Symposium Series 20 (National Technical Information Service, U.S. Department of Commerce, Springfield, Virginia).

SNYDER, W. S., FORD, M. R., WARNER, G. G., AND FISCHER, H. L., JR. (1969). "Estimates of absorbed fractions for monoenergetic photon sources uniformly distributed in various organs of a heterogenous phantom," MIRD Pamphlet 5, J. Nuc. Med. **10**, Supplement 3.

SNYDER, W. S., FORD, M. R., WARNER, G. G., AND WATSON, S. B. (1974, 1975). *A Tabulation of Dose Equivalent per Microcurie-Day for Source and Target Organs of an Adult for Various Radionuclides*, Vol. 1 (1974), Vol. 2 (1975), Oak Ridge National Laboratory Doc. ORNL-5000 (Oak Ridge National Laboratory, Oak Ridge, Tennessee).

SPIERS, F. W. (1974). *A New Approach to Setting Maximum Permissible Levels of Radionuclides in Bone*, Medical Research Council Committee on Protection Against Ionizing Radiations, Report PIRC 74/3. (Medical Research Council, London).

TAMPLIN, A. R. AND COCHRAN, T. B. (1974). *Radiation Standards for Hot Particles* (National Resources Defense Council, Washington, DC).

THORNE, M. C. (1977). "Aspects of the dosimetry of x-emitting radionuclides in bone with particular emphasis on ^{226}Ra and ^{239}Pu," Phys. Med. Biol. **22**, 36.

UNSCEAR (1977). United Nations Scientific Committee on the Effects of Atomic Radiation, *Sources and Effects of Ionizing Radiation* (United Nations, New York, New York).

UNSCEAR (1982). United Nations Scientific Committee on the Effects of Atomic Radiation, *Ionizing Radiation: Sources and Biological Effects* (United Nations, New York, New York).

USAEC (1974). United States Atomic Energy Commission, *A Radiological Assessment of the Spatial Distribution of Radiation Dose from Inhaled Plutonium*, USAEC Report Wash-1320 (United States Atomic Energy Commission, Washington, D.C.).

USNRC (1975). United States Nuclear Regulatory Commission, *Reactor Safety Study: An Assessment of Accident Risks in U.S. Commercial Nuclear Power Plants*, USNRC Report NUREG-75/014 and WASH-1400 (United States Nuclear Regulatory Commission, Washington, D.C.).

The NCRP

The National Council on Radiation Protection and Measurements is a nonprofit corporation chartered by Congress in 1964 to:

1. Collect, analyze, develop, and disseminate in the public interest information and recommendations about (a) protection against radiation and (b) radiation measurements, quantities, and units, particularly those concerned with radiation protection;
2. Provide a means by which organizations concerned with the scientific and related aspects of radiation protection and of radiation quantities, units, and measurements may cooperate for effective utilization of their combined resources, and to stimulate the work of such organizations;
3. Develop basic concepts about radiation quantities, units, and measurements, about the application of these concepts, and about radiation protection;
4. Cooperate with the International Commission on Radiological Protection, the International Commission on Radiation Units and Measurements, and other national and international organizations, governmental and private, concerned with radiation quantities, units, and measurements and with radiation protection.

The Council is the successor to the unincorporated association of scientists known as the National Committee on Radiation Protection and Measurements and was formed to carry on the work begun by the Committee.

The Council is made up of the members and the participants who serve on the eighty-two scientific committees of the Council. The scientific committees, composed of experts having detailed knowledge and competence in the particular area of the committee's interest, draft proposed recommendations. These are then submitted to the full membership of the Council for careful review and approval before being published.

The following comprise the current officers and membership of the Council:

Officers

President	WARREN K. SINCLAIR
Vice President	S. JAMES ADELSTEIN

Secretary and Treasurer W. Roger Ney
Assistant Secretary Carl D. Hobelman
Assistant Treasurer James F. Berg

Members

Seymour Abrahamson
S. James Adelstein
Roy E. Albert
Peter R. Almond
Edward L. Alpen
John A. Auxier
William J. Bair
John D. Boice, Jr.
Robert L. Brent
Antone Brooks
Thomas F. Budinger
Melvin W. Carter
George W. Casarett
Randall S. Caswell
Gerald D. Dodd
Patricia W. Durbin
Joe A. Elder
Mortimer M. Elkind
Thomas S. Ely
Edward R. Epp
Jacob I. Fabrikant
R. J. Michael Fry
Robert A. Goepp
Robert O. Gorson
Arthur W. Guy
Eric J. Hall
Naomi H. Harley
John W. Healy
William R. Hendee
John M. Heslep
Seymour Jablon
Donald G. Jacobs
A. Everette James, Jr.
Bernd Kahn
James G. Kereiakes
Charles E. Land
George R. Leopold
Ray D. Lloyd
Arthur C. Lucas
Charles W. Mays
Roger O. McClellan
James McLaughlin
Barbara J. McNeil
Thomas F. Meaney
Charles B. Meinhold
Mortimer L. Mendelsohn
William E. Mills
Dade W. Moeller
A. Alan Moghissi
Robert D. Moseley, Jr.
Wesley Nyborg
Mary Ellen O'Connor
Frank L. Parker
Andrew K. Poznanski
Norman C. Rasmussen
William C. Reinig
Chester R. Richmond
James S. Robertson
Leonard A. Sagan
William J. Schull
Glenn E. Sheline
Roy E. Shore
Warren K. Sinclair
Lewis V. Spencer
John B. Storer
William L. Templeton
Roy C. Thompson
John E. Till
Arthur C. Upton
George L. Voelz
Edward W. Webster
George M. Wilkening
H. Rodney Withers

Honorary Members

Lauriston S. Taylor, *Honorary President*

Edgar C. Barnes
Victor P. Bond
Richard F. Foster
Hymer L. Friedell
Russell H. Morgan
Robert J. Nelsen

REYNOLD F. BROWN	JOHN H. HARLEY	HAROLD H. ROSSI
AUSTIN M. BRUES	LOUIS H. HEMPELMANN, JR.	WILLIAM G. RUSSELL
FREDERICK P. COWAN	PAUL C. HODGES	JOHN H. RUST
JAMES F. CROW	GEORGE V. LEROY	EUGENE L. SAENGER
MERRILL EISENBUD	WILFRID B. MANN	J. NEWELL STANNARD
ROBLEY D. EVANS	KARL Z. MORGAN	HAROLD O. WYCKOFF

Currently, the following subgroups are actively engaged in formulating recommendations:

SC-1: Basic Radiation Protection Criteria
SC-3: Medical X-Ray, Electron Beam and Gamma-Ray Protection for Energies Up to 50 MeV (Equipment Performance and Use)
SC-16: X-Ray Protection in Dental Offices
SC-18: Standards and Measurements of Radioactivity for Radiological Use
SC-38: Waste Disposal
- Task Group on Krypton-85
- Task Group on Disposal of Accident Generated Waste Water
- Task Group on Disposal of Low-Level Waste
- Task Group on the Actinides
- Task Group on Xenon
- Task Group on Definitions of Radioactive Waste Levels

SC-40: Biological Aspects of Radiation Protection Criteria
- Task Group on Atomic Bomb Survivor Dosimetry
- Subgroup on Biological Aspects of Dosimetry of Atomic Bomb Survivors

SC-43: Natural Background Radiation
SC-44: Radiation Associated with Medical Examinations
SC-45: Radiation Received by Radiation Employees
SC-46: Operational Radiation Safety
- Task Group 1 on Warning and Access Control Systems
- Task Group 2 on Uranium Mining and Milling—Radiation Safety Programs
- Task Group 3 on ALARA for Occupationally Exposed Individuals in Clinical Radiology
- Task Group 4 on Calibration of Instrumentation
- Task Group 5 on Maintaining Radiation Protection Records
- Task Group 6 on Radiation Protection for Allied Health Personnel
- Task Group 7 on Emergency Planning

SC-47: Instrumentation for the Determination of Dose Equivalent
SC-48: Apportionment of Radiation Exposure
SC-52: Conceptual Basis of Calculations of Dose Distributions
SC-53: Biological Effects and Exposure Criteria for Radiofrequency Electromagnetic Radiation
SC-54: Bioassay for Assessment of Control of Intake of Radionuclides
SC-57: Internal Emitter Standards
- Task Group 2 on Respiratory Tract Model
- Task Group 3 on General Metabolic Models

Task Group 5 on Gastrointestinal Tract Models
Task Group 6 on Bone Problems
Task Group 8 on Leukemia Risk
Task Group 9 on Lung Cancer Risk
Task Group 10 on Liver Cancer Risk
Task Group 11 on Genetic Risk
Task Group 12 on Strontium
Task Group 13 on Neptunium

SC-59: Human Radiation Exposure Experience
SC-61: Radon Measurements
SC-63: Radiation Exposure Control in a Nuclear Emergency
SC-64: Radionuclides in the Environment
Task Group 5 on Public Exposure from Nuclear Power
Task Group 6 on Screening Models
Task Group 7 on Contaminated Soil as a Source of Radiation Exposure
SC-65: Quality Assurance and Accuracy in Radiation Protection Measurements
SC-67: Biological Effects of Magnetic Fields
SC-68: Microprocessors in Dosimetry
SC-69: Efficacy of Radiographic Procedures
SC-70: Quality Assurance and Measurement in Diagnostic Radiology
SC-71: Radiation Exposure and Potentially Related Injury
SC-72: Radiation Protection in Mammography
SC-74: Radiation Received in the Decontamination of Nuclear Facilities
SC-75: Guidance on Radiation Received in Space Activities
SC-76: Effects of Radiation on the Embryo-Fetus
SC-77: Guidance on Occupational and Public Exposure Resulting from Diagnostic Nuclear Medicine Procedures
SC-78: Practical Guidance on the Evaluation of Human Exposures to Radiofrequency Radiation
SC-79: Extremely Low-Frequency Electric and Magnetic Fields
SC-80: Radiation Biology of the Skin (Beta-Ray Dosimetry)
SC-81: Assessment of Exposure from Therapy
SC-82: Control of Indoor Radon

Committee on Public Education
Committee on Public Relations
Ad Hoc Committee on Comparison of Radiation Exposures
Study Group on Comparative Risk
Task Group on Comparative Carcinogenicity of Pollutant Chemicals
Task Force on Occupational Exposure Levels
Ad Hoc Group on Model Used for Assessing Transport of Low-Level Radioactive Waste
Ad Hoc Group on Medical Evaluation of Radiation Workers
Task Group on Ocean Dumping of Radioactive Waste

In recognition of its responsibility to facilitate and stimulate cooperation among organizations concerned with the scientific and related aspects of radiation protection and measurement, the Council has created a category of NCRP Collaborating Organizations. Organizations or groups of organizations that are national or international in scope and are concerned with scientific problems involving radiation quantities, units, measurements, and effects, or radiation protection may be admitted to collaborating status by the Council. The present Collaborating Organizations with which the NCRP maintains liaison are as follows:

American Academy of Dermatology
American Association of Physicists in Medicine
American College of Nuclear Physicians
American College of Radiology
American Dental Association
American Industrial Hygiene Association
American Institute of Ultrasound in Medicine
American Insurance Association
American Medical Association
American Nuclear Society
American Occupational Medical Association
American Podiatric Medical Association
American Public Health Association
American Radium Society
American Roentgen Ray Society
American Society of Radiologic Technologists
American Society for Therapeutic Radiology and Oncology
Association of University Radiologists
Atomic Industrial Forum
Bioelectromagnetics Society
College of American Pathologists
Federal Communications Commission
Federal Emergency Management Agency
Genetics Society of America
Health Physics Society
National Bureau of Standards
National Electrical Manufacturers Association
Radiation Research Society
Radiological Society of North America
Society of Nuclear Medicine
United States Air Force
United States Army
United States Department of Energy
United States Department of Housing and Urban Development
United States Department of Labor
United States Environmental Protection Agency
United States Navy
United States Nuclear Regulatory Commission
United States Public Health Service

The NCRP has found its relationships with these organizations to be extremely valuable to continued progress in its program.

Another aspect of the cooperative efforts of the NCRP relates to the special liaison relationships established with various governmental organizations that have an interest in radiation protection and measurements. This liaison relationship provides: (1) an opportunity for participating organizations to designate an individual to provide liaison between the organization and the NCRP; (2) that the individual designated will receive copies of draft NCRP reports (at the time that these are submitted to the members of the Council) with an invitation to comment, but not vote; and (3) that new NCRP efforts might be discussed with liaison individuals as appropriate, so that they might have an opportunity to make suggestions on new studies and related matters. The following organizations participate in the special liaison program:

Commission of the European Communities
Commisariat a l'Energie Atomique (France)
Defense Nuclear Agency
Federal Emergency Management Agency
Japan Radiation Council
National Bureau of Standards
National Radiological Protection Board (United Kingdom)
National Research Council (Canada)
Office of Science and Technology Policy
Office of Technology Assessment
United States Air Force
United States Army
United States Coast Guard
United States Department of Energy
United States Department of Health and Human Services
United States Department of Labor
United States Department of Transportation
United States Environmental Protection Agency
United States Navy
United States Nuclear Regulatory Commission

The NCRP values highly the participation of these organizations in the liaison program.

The Council's activities are made possible by the voluntary contribution of time and effort by its members and participants and the generous support of the following organizations:

Alfred P. Sloan Foundation
Alliance of American Insurers
American Academy of Dental Radiology
American Academy of Dermatology
American Association of Physicists in Medicine
American College of Nuclear Physicians
American College of Radiology
American College of Radiology Foundation
American Dental Association
American Hospital Radiology Administrators

American Industrial Hygiene Association
American Insurance Association
American Medical Association
American Nuclear Society
American Occupational Medical Association
American Osteopathic College of Radiology
American Podiatric Medical Association
American Public Health Association
American Radium Society
American Roentgen Ray Society
American Society of Radiologic Technologists
American Society for Therapeutic Radiology and Oncology
American Veterinary Medical Association
American Veterinary Radiology Society
Association of University Radiologists
Atomic Industrial Forum
Battelle Memorial Institute
Center for Devices and Radiological Health
College of American Pathologists
Commonwealth of Pennsylvania
Defense Nuclear Agency
Edison Electric Institute
Edward Mallinckrodt, Jr. Foundation
Electric Power Research Institute
Federal Emergency Management Agency
Florida Institute of Phosphate Research
Genetics Society of America
Health Physics Society
James Picker Foundation
National Aeronautics and Space Administration
National Association of Photographic Manufacturers
National Bureau of Standards
National Cancer Institute
National Electrical Manufacturers Association
Radiation Research Society
Radiological Society of North America
Society of Nuclear Medicine
United States Department of Energy
United States Department of Labor
United States Environmental Protection Agency
United States Navy
United States Nuclear Regulatory Commission

To all of these organizations the Council expresses its profound appreciation for their support.

Initial funds for publication of NCRP reports were provided by a grant from the James Picker Foundation and for this the Council wishes to express its deep appreciation.

The NCRP seeks to promulgate information and recommendations based on leading scientific judgment on matters of radiation protection and measurement and to foster cooperation among organizations concerned with these matters. These efforts are intended to serve the public interest and the Council welcomes comments and suggestions on its reports or activities from those interested in its work.

NCRP Publications

NCRP publications are distributed by the NCRP Publications' office. Information on prices and how to order may be obtained by directing an inquiry to:

NCRP Publications
7910 Woodmont Ave., Suite 1016
Bethesda, MD 20814

The currently available publications are listed below.

Proceedings of the Annual Meeting

No.	Title
1	*Perceptions of Risk*, Proceedings of the Fifteenth Annual Meeting, Held on March 14–15, 1979 (Including Taylor Lecture No. 3) (1980)
2	*Quantitative Risk in Standards Setting*, Proceedings of the Sixteenth Annual Meeting Held on April 2–3, 1980 (Including Taylor Lecture No. 4) (1981)
3	*Critical Issues in Setting Radiation Dose Limits*, Proceedings of the Seventeenth Annual Meeting, Held on April 8–9, 1981 (Including Taylor Lecture No. 5) (1982)
4	*Radiation Protection and New Medical Diagnostic Procedures*, Proceedings of the Eighteenth Annual Meeting, Held on April 6–7, 1982 (Including Taylor Lecture No. 6) (1983)
5	*Environmental Radioactivity*, Proceedings of the Nineteenth Annual Meeting, Held on April 6–7, 1983 (Including Taylor Lecture No. 7) (1984)
6	*Some Issues Important in Developing Basic Radiation Protection Recommendations*, Proceedings of the Twentieth Annual Meeting, Held on April 4–5, 1984 (Including Taylor Lecture No. 8) (1985)

Symposium Proceedings

The Control of Exposure of the Public to Ionizing Radiation in the Event of Accident or Attack, Proceedings of a Symposium held April 27–29, 1981 (1982)

Lauriston S. Taylor Lectures

No.	Title and Author
1	*The Squares of the Natural Numbers in Radiation Protection* by Herbert M. Parker (1977)
2	*Why be Quantitative About Radiation Risk Estimates?* by Sir Edward Pochin (1978)
3	*Radiation Protection—Concepts and Trade Offs* by Hymer L. Friedell (1979) [Available also in *Perceptions of Risk*, see above]
4	*From "Quantity of Radiation" and "Dose" to "Exposure" and "Absorbed Dose"—An Historical Review* by Harold O. Wyckoff (1980) [Available also in *Quantitative Risks in Standards Setting*, see above]
5	*How Well Can We Assess Genetic Risk? Not Very* by James F. Crow (1981) [Available also in *Critical Issues in Setting Radiation Dose Limits*, see above]
6	*Ethics, Trade-offs and Medical Radiation* by Eugene L. Saenger (1982) [Available also in *Radiation Protection and New Medical Diagnostic Approaches*, see above]
7	*The Human Environment-Past, Present and Future* by Merril Eisenbud (1983) [Available also in *Environmental Radioactivity*, see above]
8	*Limitation and Assessment in Radiation Protection* by Harald H. Rossi (1984) [Available also in 1985 in *Some Issues Important in Developing Basic Radiation Protection Recommendations*, see above]
9	*Truth (and Beauty) in Radiation Measurement* by John H. Harley (1985) [Available also in 1985 in *Radioactive Waste*]

NCRP Reports

No.	Title
8	*Control and Removal of Radioactive Contamination in Laboratories* (1951)
9	*Recommendations for Waste Disposal of Phosphorus-32 and Iodine-131 for Medical Users* (1951)
12	*Recommendations for the Disposal of Carbon-14 Wastes* (1953)
16	*Radioactive Waste Disposal in the Ocean* (1954)
22	*Maximum Permissible Body Burdens and Maximum Per-*

missible Concentrations of Radionuclides in Air and in Water for Occupational Exposure* (1959) [Includes Addendum 1 issued in August 1963]

23 *Measurement of Neutron Flux and Spectra for Physical and Biological Applications* (1960)

25 *Measurement of Absorbed Dose of Neutrons and Mixtures of Neutrons and Gamma Rays* (1961)

27 *Stopping Powers for Use with Cavity Chambers* (1961)

30 *Safe Handling of Radioactive Materials* (1964)

32 *Radiation Protection in Educational Institutions* (1966)

33 *Medical X-Ray and Gamma-Ray Protection for Energies Up to 10 MeV—Equipment Design and Use* (1968)

35 *Dental X-Ray Protection* (1970)

36 *Radiation Protection in Veterinary Medicine* (1970)

37 *Precautions in the Management of Patients Who Have Received Therapeutic Amounts of Radionuclides* (1970)

38 *Protection against Neutron Radiation* (1971)

39 *Basic Radiation Protection Criteria* (1971)

40 *Protection Against Radiation from Brachytherapy Sources* (1972)

41 *Specification of Gamma-Ray Brachytherapy Sources* (1974)

42 *Radiological Factors Affecting Decision-Making in a Nuclear Attack* (1974)

43 *Review of the Current State of Radiation Protection Philosophy* (1975)

44 *Krypton-85 in the Atmosphere—Accumulation, Biological Significance, and Control Technology* (1975)

45 *Natural Background Radiation in the United States* (1975)

46 *Alpha-Emitting Particles in Lungs* (1975)

47 *Tritium Measurement Techniques* (1976)

48 *Radiation Protection for Medical and Allied Health Personnel* (1976)

49 *Structural Shielding Design and Evaluation for Medical Use of X Rays and Gamma Rays of Energies Up to 10 MeV* (1976)

50 *Environmental Radiation Measurements* (1976)

51 *Radiation Protection Design Guidelines for 0.1–100 MeV Particle Accelerator Facilities* (1977)

52 *Cesium-137 From the Environment to Man: Metabolism and Dose* (1977)

53 *Review of NCRP Radiation Dose Limit for Embryo and Fetus in Occuptionally Exposed Women* (1977)

54 *Medical Radiation Exposure of Pregnant and Potentially*

Pregnant Women (1977)

55 *Protection of the Thyroid Gland in the Event of Releases of Radioiodine* (1977)

56 *Radiation Exposure From Consumer Products and Miscellaneous Sources* (1977)

57 *Instrumentation and Monitoring Methods for Radiation Protection* (1978)

58 *A Handbook of Radioactivity Measurements Procedures*, 2nd Ed. (1985)

59 *Operational Radiation Safety Program* (1978)

60 *Physical, Chemical, and Biological Properties of Radiocerium Relevant to Radiation Protection Guidelines* (1978)

61 *Radiation Safety Training Criteria for Industrial Radiography* (1978)

62 *Tritium in the Environment* (1979)

63 *Tritium and Other Radionuclide Labeled Organic Compounds Incorporated in Genetic Material* (1979)

64 *Influence of Dose and Its Distribution in Time on Dose-Response Relationships for Low-LET Radiations* (1980)

65 *Management of Persons Accidentally Contaminated with Radionuclides* (1980)

66 *Mammography* (1980)

67 *Radiofrequency Electromagnetic Fields—Properties, Quantities and Units, Biophysical Interaction, and Measurements* (1981)

68 *Radiation Protection in Pediatric Radiology* (1981)

69 *Dosimetry of X-Ray and Gamma-Ray Beams for Radiation Therapy in the Energy Range 10 keV to 50 MeV* (1981)

70 *Nuclear Medicine-Factors Influencing the Choice and Use of Radionuclides in Diagnosis and Therapy* (1982)

71 *Operational Radiation Safety—Training* (1983)

72 *Radiation Protection and Measurement for Low Voltage Neutron Generators* (1983)

73 *Protection in Nuclear Medicine and Ultrasound Diagnostic Procedures in Children* (1983)

74 *Biological Effects of Ultrasound: Mechanisms and Clinical Implications* (1983)

75 *Iodine-129: Evaluation of Releases from Nuclear Power Generation* (1983)

76 *Radiological Assessment: Predicting the Transport, Bioaccumulation, and Uptake by Man of Radionuclides Released to the Environment* (1984)

77 *Exposures from the Uranium Series with Emphasis on Radon and its Daughters* (1984)

78 *Evaluation of Occupational and Environmental Exposures to Radon and Radon Daughters in the United States* (1984)
79 *Neutron Contamination from Medical Electron Accelerators* (1984)
80 *Induction of Thyroid Cancer by Ionizing Radiation* (1985)
81 *Carbon-14 in the Environment* (1985)
82 *SI Units in Radiation Protection and Measurements* (1985)
83 *The Experimental Basis for Absorbed Dose-Calculations in Medical uses of Radionuclides* (1985)
84 *General Concepts for the Dosimetry of Internally Deposited Radionuclides* (1985)

Binders for NCRP Reports are available. Two sizes make it possible to collect into small binders the "old series" of reports (NCRP Reports Nos. 8–30) and into large binders the more recent publications (NCRP Reports Nos. 32–83). Each binder will accommodate from five to seven reports. The binders carry the identification "NCRP Reports" and come with label holders which permit the user to attach labels showing the reports contained in each binder.

The following bound sets of NCRP Reports are also available:

Volume I. NCRP Reports Nos. 8, 9, 12, 16, 22
Volume II. NCRP Reports Nos. 23, 25, 27, 30
Volume III. NCRP Reports Nos. 32, 33, 35, 36, 37
Volume IV. NCRP Reports Nos. 38, 39, 40, 41
Volume V. NCRP Reports Nos. 42, 43, 44, 45, 46
Volume VI. NCRP Reports Nos. 47, 48, 49, 50, 51
Volume VII. NCRP Reports Nos. 52, 53, 54, 55, 56, 57
Volume VIII. NCRP Report No. 58
Volume IX. NCRP Reports Nos. 59, 60, 61, 62, 63
Volume X. NCRP Reports Nos. 64, 65, 66, 67
Volume XI. NCRP Reports Nos. 68, 69, 70, 71, 72
Volume XII. NCRP Reports Nos. 73, 74, 75, 76

(Titles of the individual reports contained in each volume are given above).

The following NCRP Reports are now superseded and/or out of print:

No.	Title
1	*X-Ray Protection* (1931). [Superseded by NCRP Report No. 3]

2 *Radium Protection* (1934). [Superseded by NCRP Report No. 4]

3 *X-Ray Protection* (1936). [Superseded by NCRP Report No. 6]

4 *Radium Protection* (1938). [Superseded by NCRP Report No. 13]

5 *Safe Handling of Radioactive Luminous Compounds* (1941). [Out of Print]

6 *Medical X-Ray Protection Up to Two Million Volts* (1949). [Superseded by NCRP Report No. 18]

7 *Safe Handling of Radioactive Isotopes* (1949). [Superseded by NCRP Report No. 30]

10 *Radiological Monitoring Methods and Instruments* (1952). [Superseded by NCRP Report No. 57]

11 *Maximum Permissible Amounts of Radioisotopes in the Human Body and Maximum Permissible Concentrations in Air and Water* (1953). [Superseded by NCRP Report No. 22]

13 *Protection Against Radiations from Radium, Cobalt-60 and Cesium-137* (1954). [Superseded by NCRP Report No. 24]

14 *Protection Against Betatron—Synchrotron Radiations Up to 100 Million Electron Volts* (1954). [Superseded by NCRP Report No. 51]

15 *Safe Handling of Cadavers Containing Radioactive Isotopes* (1953). [Superseded by NCRP Report No. 21]

17 *Permissible Dose from External Sources of Ionizing Radiation* (1954) including *Maximum Permissible Exposure to Man, Addendum to National Bureau of Standards Handbook 59* (1958). [Superseded by NCRP Report No. 39]

18 *X-Ray Protection* (1955). [Superseded by NCRP Report No. 26

19 *Regulation of Radiation Exposure by Legislative Means* (1955). [Out of print]

20 *Protection Against Neutron Radiation Up to 30 Million Electron Volts* (1957). [Superseded by NCRP Report No. 38]

21 *Safe Handling of Bodies Containing Radioactive Isotopes* (1958). [Superseded by NCRP Report No. 37]

24 *Protection Against Radiations from Sealed Gamma Sources* (1960). [Superseded by NCRP Report Nos. 33, 34, and 40]

26 *Medical X-Ray Protection Up to Three Million Volts* (1961). [Superseded by NCRP Report Nos. 33, 34, 35, and 36]

28 *A Manual of Radioactivity Procedures* (1961). [Superseded by NCRP Report No. 58]

29 *Exposure to Radiation in an Emergency* (1962). [Superseded by NCRP Report No. 42]

31 *Shielding for High Energy Electron Accelerator Installations* (1964). [Superseded by NCRP Report No. 51]

34 *Medical X-Ray and Gamma-Ray Protection for Energies Up to 10 MeV—Structural Shielding Design and Evaluation* (1970). [Superseded by NCRP Report No. 49]

Other Documents

The following documents of the NCRP were published outside of the NCRP Reports series:

"Blood Counts, Statement of the National Committee on Radiation Protection," Radiology 63, 428 (1954)

"Statements on Maximum Permissible Dose from Television Receivers and Maximum Permissible Dose to the Skin of the Whole Body," Am. J. Roentgenol., Radium Ther. and Nucl. Med. 84, 152 (1960) and Radiology 75, 122 (1960)

X-Ray Protection Standards for Home Television Receivers, Interim Statement of the National Council on Radiation Protection and Measurements (National Council on Radiation Protection and Measurements, Washington, 1968)

Specification of Units of Natural Uranium and Natural Thorium (National Council on Radiation Protection and Measurements, Washington, 1973)

NCRP Statement on Dose Limit for Neutrons (National Council on Radiation Protection and Measurements, Washington, 1980)

Krypton-85 in the Atmosphere—With Specific Reference to the Public Health Significance of the Proposed Controlled Release at Three Mile Island (National Council on Radiation Protection and Measurements, Washington, 1980)

Preliminary Evaluation of Criteria For the Disposal of Transuranic Contaminated Waste (National Council on Radiation Protection and Measurements, Bethesda, Maryland, 1982)

Control of Air Emissions of Radionuclides (National Council on Radiation Protection and Measurements, Bethesda, Maryland, 1984)

Copies of the statements published in journals may be consulted in libraries. A limited number of copies of the remaining documents listed above are available for distribution by NCRP Publications.

INDEX